Lecture Notes in Mechanical Engineering

T0215893

About this Series

Lecture Notes in Mechanical Engineering (LNME) publishes the latest developments in Mechanical Engineering—quickly, informally and with high quality. Original research reported in proceedings and post-proceedings represents the core of LNME. Also considered for publication are monographs, contributed volumes and lecture notes of exceptionally high quality and interest. Volumes published in LNME embrace all aspects, subfields and new challenges of mechanical engineering. Topics in the series include:

- Engineering Design
- Machinery and Machine Elements
- Mechanical Structures and Stress Analysis
- Automotive Engineering
- Engine Technology
- Aerospace Technology and Astronautics
- Nanotechnology and Microengineering
- Control, Robotics, Mechatronics
- MEMS
- Theoretical and Applied Mechanics
- Dynamical Systems, Control
- Fluid Mechanics
- Engineering Thermodynamics, Heat and Mass Transfer
- Manufacturing
- Precision Engineering, Instrumentation, Measurement
- Materials Engineering
- Tribology and Surface Technology

More information about this series at http://www.springer.com/series/11236

Alexander Evgrafov
Editor

Advances in Mechanical Engineering

Selected Contributions from the Conference
"Modern Engineering: Science
and Education", Saint Petersburg,
Russia, June 20–21, 2013

 Springer

Editor
Alexander Evgrafov
Saint Petersburg State Polytechnical
    University
Saint Petersburg
Russia

ISSN 2195-4356   ISSN 2195-4364   (electronic)
Lecture Notes in Mechanical Engineering
ISBN 978-3-319-15683-5   ISBN 978-3-319-15684-2   (eBook)
DOI 10.1007/978-3-319-15684-2

Library of Congress Control Number: 2015932506

Springer Cham Heidelberg New York Dordrecht London

Springer International Publishing AG Switzerland is part of Springer Science+Business Media (www.springer.com)

Preface

The "Modern Engineering: Science and Education" (MESE) conference was initially organized by the Mechanical Engineering Department of Saint Petersburg State Polytechnic University in June 2011 in St. Petersburg (Russia). It was envisioned as a forum in which to bring together scientists, university professors, graduate students, and mechanical engineers, presenting new science, technology, and engineering ideas and achievements.

The idea of holding such a forum proved to be highly relevant. Moreover, both location and timing of the conference were quite appealing. Late June is a wonderful and romantic season in St. Petersburg—one of the most beautiful cities, located on the Neva river banks, and surrounded by charming greenbelts. The conference attracted many participants, working in various fields of engineering: design, mechanics, materials, etc. The success of the conference inspired the organizers to turn the conference into an annual event.

The third conference, MESE 2013, attracted 150 presentations and covered topics ranging from mechanics of machines, materials engineering, structural strength, and tribological behavior to transport technologies, machinery quality and innovations, in addition to dynamics of machines, walking mechanisms, and computational methods. All presenters contributed greatly to the success of the conference. However, for the purposes of this book only 16 reports, authored by research groups representing various universities and institutes, were selected for inclusion.

I am particularly grateful to the authors for their contributions and all the participating experts for their valuable advice. Furthermore, I would like to thank the staff and management of the University for their cooperation and support, and especially, all members of the program committee and the organizing committee for their work in preparing and organizing the conference.

Last, but not least, I would like to thank Springer for its professional assistance and particularly Mr. Pierpaolo Riva who supported this publication.

Saint Petersburg   Alexander Evgrafov

Contents

1   **Rotary Forging of Hollow Components with Flanges**   1
     Leonid B. Aksenov and Sergey N. Kunkin

2   **On the Research Technique of Mesogene Lubrication Layer
     Optical Properties**   7
     Elena V. Berezina, Alexej V. Volkov, Vladimir A. Godlevskiy,
     Anton G. Zheleznov and Dmitry S. Fomichev

3   **Auto-Ignition Problem Titanium of Oxygen and Possible
     Ways of Solving**   13
     V.I. Bolobov

4   **Control of Biped Walking Robot Using Equations
     of the Inverted Pendulum**   23
     Valeriy Tereshin and Anastasiya Borina

5   **An Efficient Model for the Orthogonal Packing Problem**   33
     Vladislav A. Chekanin and Alexander V. Chekanin

6   **Designing a Power Converter with an Adaptive Control
     System for Ultrasonic Processing Units**   39
     Vladimir I. Diagilev, Valery A. Kokovin and Saygid U. Uvaysov

7   **Features of Multicomponent Saturation Alloyed by Steels**   49
     Sergey G. Ivanov, Irina A. Garmaeva, Michael A. Guriev,
     Alexey M. Guriev and Michael D. Starostenkov

8   **New Modelling and Calculation Methods for Vibrating
     Screens and Separators**   55
     Kirill S. Ivanov and Leonid A. Vaisberg

9   Remote Monitoring Systems for Quality Management
     Metal Pouring   63
     Vladimir F. Minakov and Tatyana E. Minakova

10 Acoustic Emission Monitoring of Leaks   73
     Evgeny Nefedyev

11 Dynamic Model of the Two-Mass Active Vibroprotective
     System   85
     Mikhail J. Platovskikh

12 Structural and Phase Transformation in Material of Blades
     of Steam Turbines from Titanium Alloy After Technological
     Treatment   93
     Margarita A. Skotnikova, Galina V. Tsvetkova,
     Aleksandra A. Lanina, Nikolay A. Krylov
     and Galina V. Ivanova

13 Conflicts in Product Development and Machining Time
     Estimation at Early Design Stages   103
     Dmitry I. Troitsky

14 Vibrations Excitation in Cyclic Mechanisms Due to Energy
     Generated in Nonstationary Constraints   117
     Iosif I. Vulfson

15 Dynamics, Critical Speeds and Balancing of Thermoelastic
     Rotors   129
     Vladimir V. Yeliseyev

16 Influence of Plastic Deformation on Fatigue of Titanium
     Alloys   137
     Vladimir A. Zhukov

Chapter 1
Rotary Forging of Hollow Components with Flanges

Leonid B. Aksenov and Sergey N. Kunkin

Abstract This paper is about development and testing of cold rotary forging technology for manufacturing of hollow shaft flanges. Rotary forging mills with rotated die and conical roll were used for forming. Pre-machined blanks were made from pieces of welded or extruded pipes from carbon steel with 0.2 % of carbon. Design of tool-sets, sketches of blank and forged parts are presented.

Keywords Rotary forging · Shaft flanges · Conical roll

Introduction

Ax symmetric parts with flanges at some distance from the tube face (Fig. 1.1) are widely used by industry for the production of various machines. Technologies for their production differ greatly. One such process is radial extrusion of tube blanks in hot condition. Also technology of hot die forging at maxi-presses and forging machines is very popular. In most cases, the coefficient of use of metal is low, and forged parts require a large amount of machining.

Manufacturers of pipes and pipe fittings in various industries would like to make flanges directly from its main product—pipes. This is possible to realize with technology of rotary forging [1]. The rotary forging technology is intended for manufacture of ax symmetric parts from bar or pipe blanks. This technology is representative of processes with local deformation [2]. Only part of the blank is in contact with a deforming tool. It reduces the contact area, contact stresses and the required forming force. Rotary forging is attractive technology in the field of metal

L.B. Aksenov (✉) · S.N. Kunkin
St. Petersburg State Polytechnic University, St. Petersburg, Russia
e-mail: l_axenov@mail.spbstu.ru

S.N. Kunkin
e-mail: kunkin@spbstu.ru

© Springer International Publishing Switzerland 2015
A. Evgrafov (ed.), *Advances in Mechanical Engineering*, Lecture Notes
in Mechanical Engineering, DOI 10.1007/978-3-319-15684-2_1

Fig. 1.1 Cross section of
hollow shaft with flange

forming, because it has many advantages over any other processes: smaller
deformation force, longer die life, less investment in equipment [3].

The process of cold rotary forging provides additional advantages. It does not
require heating, provides high accuracy and good quality of the formed surfaces [4].
Obviously, in a cold forging, technological force will be higher than at a hot one
with lower plasticity of metal. That is why this technology needs more stringent
requirements in their analysis [5].

Experimental Procedure

Investigation has been done in the direction of development of rotary forming in
which a rotating blank is deformed by forging a roll moving along the axis of the
blank [6]. The rotary motion is provided by a motor driving the lower turntable
while the upper die rotates as a follower through the blank.

The blank is placed in the die with a radial clearance of 0.3 mm. It is not
necessary to fix a blank during the forming. On the first stage of forming, the
rotation of the blank takes place due to friction force between the face of the blank
and forming roll. Then the blank is pressed to the die and this provides a reliable
transmission of torque. During the rotary forging, the blank is shaped in the space
that is formed by the die, mandrel and deforming tool—forging roll. The tool set
(die and forging roll) is manufactured from die steel with heat treatment for
hardness HRC = 56–63 [6].

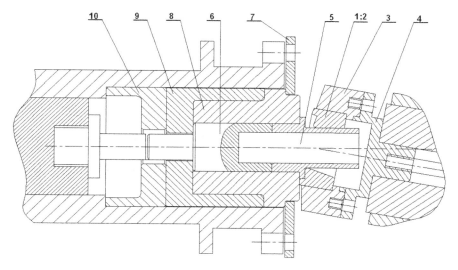

Fig. 1.2 Tool set for axial rotary forging of hollow shaft flanges: *1, 2* forging roll, *3* mandrel, *4* shank, *5* inner mandrel, *6* stop, *7* top, *8* die, *9* cup, *10* base

This experimental work investigated the possibility of manufacturing ax symmetric hollow parts with outer flange. A tool set for rotary forging (Fig. 1.2) was designed and manufactured for a horizontal type of mill [7].

Rotary forging was realized with the following main process parameters:

- angle of inclination of the conical roll—10°;
- speed of rotation of a spindle of the machine—130 rev/min;
- feed of forging roll—0.3 mm/rev in the beginning of the process and about 0.05 mm/rev at the end of the forming;
- lubricant—machine oil;
- duration of rotary forging (forming time)—20–25s.

Pre-machined pieces of pipe were used as blanks for rotary forging. Material—steel 20, Russian standard: GOST 1050-88. Outer diameter—57 mm and internal diameter—42 mm as at pipe billet. The initial hardness of steel 20 was HV 140…145.

Results and Discussion

Sketches of a pre-machined steel blank for further rotary forging and a rotary forged part are presented in Fig. 1.3. Final dimensions of rotary forged parts are shown in Table 1.1. Dimensions of the blank are determined by dimensions of the manufactured part and volume of the metal required for flange forming.

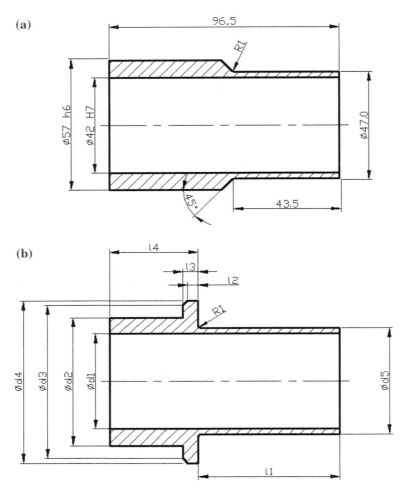

Fig. 1.3 Sketches of pre-machined blank (**a**) and rotary forged part (**b**)

Table 1.1 The main dimensions of rotary forged parts

d1 (mm)	d2 (mm)	d3 (mm)	d4 (mm)	d5 (mm)	l1 (mm)	l2 (mm)	l3 (mm)	l4 (mm)
42.0	57.0	68.5	72.0	47.0	62.2	5.5	6.5	37.4

For guarantee of full filling of the flange of the parts from steel 20 required the forging force to be 400...500 kN. Research also found that the use of a pre-shaped blank and plasticity resource of material (steel 20) were sufficient for rotary forging the parts in one operation and to obtain a flange with a diameter of 72 mm without any cracks (Fig. 1.4).

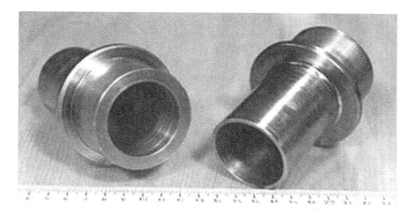

Fig. 1.4 Rotary forged parts from tube-blanks

Resume

1. A stable technological process for manufacturing of hollow shafts with flange diameter 72 mm, requires that a rotary forging mill have a capacity (forging force) of 500 kN.
2. Manufacturing of parts with a constant inner diameter can be made in one operation of rotary forging by using pre-shaped blanks.
3. Pre-shaped blanks for rotary forging processes can be made from welded or extruded pipes by machining or rotary reduction.

Acknowledgments This research was sponsored by SMS Eumuco GmbH, Wagner Banning, Ring Rolling Division (Germany).

References

1. Standring PM (1999) The significance of nutation angle in rotary forging. Advance technology of plasticity, vol III. In: Proceedings of the 6th ICTP 19–24 Sept 1999
2. Groche P, Fritsche D, Tekkaya EA, Allwood JM, Hirt G, Neugebauer R (2007) Incremental bulk metal forming. Ann CIRP 56:635–656
3. Xinghui Han, Lin Hua (2009) Comparison between cold rotary forging and conventional forging. J Mech Sci Technol 23:2668–2678
4. Nowak J, Madej L, Ziolkiewicz S, Plewinski A, Grosman F, Pietrzyk M (2008) Recent development in orbital forging technology. Int J Mater Form (Suppl 1):387–390
5. Deng XB, Hua L, Han XH et al (2011) Numerical and experimental investigation of cold rotary forging of a 20CrMnTi alloy spur bevel gear. Mater Des 32:1376–1389
6. Kunkin SN, Aksenov LB (2013) Technology of manufacturing by rotary forging of ax symmetric components with flange. Mod Mach Build Sci Educ 3:858–866 (in Russian)
7. Aksenov LB, Kunkin SN, Elkin NM (2011) Face rotary forging of flange components for tube connectors. Metalworking 63(3):31–36 (Metalloobrabotka, in Russian)

Chapter 2
On the Research Technique of Mesogene Lubrication Layer Optical Properties

Elena V. Berezina, Alexej V. Volkov, Vladimir A. Godlevskiy,
Anton G. Zheleznov and Dmitry S. Fomichev

Abstract The principles of polarizational tribometry and its instrument realization
are described. Models of thin lubricating layer rheology are proposed and the
possibilities of recording for an object's optical anisotropy in these models are
compared.

Keywords Lubricating materials · Lubrication layer · Polarizing tribometer ·
Anisotropy · Mesogen · Double refraction

Introduction

Polarizing microscopy is the most widespread method of measuring the phase
conditions of substances research, including mesomorphic structures. It allows
observation of bulk structuring effects also in solutions. It provides an opportunity
to also use the given technique to determine the phase state estimation of investigated lubricants. Presence of double refraction structures in lubricant volume allows
one to assume the higher aptitude of boundary lubricant layer self-organizing on a
solid surface [1, 2].

E.V. Berezina (✉) · A.V. Volkov · V.A. Godlevskiy · A.G. Zheleznov · D.S. Fomichev
Ivanovo State University, Ivanovo, Russia
e-mail: elena_berezina@mail.ru

A.V. Volkov
e-mail: volkz@ivanovo.ac.ru

V.A. Godlevskiy
e-mail: godl@yandex.ru

A.G. Zheleznov
e-mail: antoniron@smtp.ru

D.S. Fomichev
e-mail: dsf81@mail.ru

© Springer International Publishing Switzerland 2015
A. Evgrafov (ed.), *Advances in Mechanical Engineering*, Lecture Notes
in Mechanical Engineering, DOI 10.1007/978-3-319-15684-2_2

There is little data in the literature about attempts of application of polarizing-optical techniques to lubricant layers of mesogene nature, though this approach could essentially clarify the situation [3]. However, it is necessary to overcome the difficulties both theoretical, and experimental in this direction.

The orientation effects at molecular level, however, do occur but not so much owing to the proximity of superficial atoms but mostly due to the result of near-surface shift. Despite the rational evidence of the described mechanism, experimental proofs of its existence are obviously not sufficient. Within the last several decades many speculative models showing supra-molecular arrangement of lubrication layers nearest to a solid surface have been put forward.

Especially it concerned surface-active substances, and in general tribo-active components of anisometric molecular structure. The structural models have been presented both as mono- and multilayer structures. Only during the last few years the possibility has appeared to check the adequacy of such models with the help of molecular modeling methods, to estimate possibility of existence of other kinds of molecular packing.

To reveal how molecular orientation effects influence rheological properties of a lubrication layer, "polarizing tribometer" types of devices have been offered. These devices use the interaction of polarized light rays with lubrication substances in which the shear occurs. Attempts to get mesogene lubricant layer images received with the help of polarized light during translational motion of friction pair [4] or at a relative movement of friction surfaces along the elliptic trajectory [5] are known.

The Experimental Set

The installation which is based on the described above principle and realizing the rotary scheme "disk–disk" where disks co-operate with their flat surfaces was created by the authors of the present article. The objective of our experiment was simultaneous registration of friction force moment and the optical signal reflecting anisotropy of lubricant [6]. The tribometrical scheme based on rotary movement of friction surfaces, creating stable conditions of shear process for display of optical effects in a lubricant layer, was realized in the device.

The friction pair represents two optical windows of disk-like form. The bottom disk is motionless, the top one is put into rotation. The sample of lubricant is located between the glasses; the working backlash is regulated by the micrometric device.

Based on the drive kinematic scheme and surface properties of the glasses, it was possible to provide a minimum gap of a friction pair not exceeding 30 mcm. We used laser as a source of light in the optical part of the device. As a laser beam by its nature is polarized, adding the second rotary polarizer into the system allowed to "extinguish" the beam at the beginning of the experiment (when the anisometry caused by the shear has not yet caused the anisometry of the lubricant layer structure) similarly to the way it is done for liquid crystals mesomorphism research. At the beginning of a friction pair relative movement, thus, there appeared the orientation of the lubricant's mesogene molecules, and at an exit of the optical system there appeared a signal reflecting the supra-molecular orientation processes.

Let's try to formulate some considerations concerning what information on the condition of the lubrication layer we can get with the help of the technique using polarized light. It is possible to consider, that the anisotropy parameter of mesogene lubrication layer material is proportional to the intensity of optical transmission which, in turn, depends on average shear stress or deformation or on the rotating moment in case of rotary relative movement of friction surfaces. As far as in such an experiment it is impossible to measure the rotary moment in that local area where direct measurement of light transmission occurs, it is necessary to accept that the given moment is proportional to some average moment which is measured directly by the tribometer.

Theory

The liquid movement at the established current is described by the Navier–Stokes equation in cylindrical coordinates [7]. From the equation record we believe that, for the radius of working disk r on which we make measurements, it is essential that more than the average thickness of working backlash h be:

$$\Delta v_\varphi = \frac{v_\varphi}{r^2} \text{ or } \frac{1}{r}\frac{\partial}{\partial r}\left(r\frac{\partial v_\varphi}{\partial r}\right) + \frac{\partial^2 v_\varphi}{\partial z^2} = \frac{v_\varphi}{r^2} \tag{2.1}$$

$$\text{Boundary conditions: } z = 0: \ v_\varphi = 0$$
$$z = h: \ v_\varphi = \omega r$$

where h—thickness of interface backlash, ω—angular velocity of relative friction pair rotation; r—radius on which to make measurements, v_φ—nonzero component of linear speed.

Shear stress in the given system can be written down as:

$$\begin{cases} \sigma_{\varphi z} = \eta \frac{\partial v_\varphi}{\partial z} \\ \sigma_{r\varphi} = \eta\left(\frac{\partial v_\varphi}{\partial r} - \frac{v_\varphi}{r}\right) \end{cases}, \tag{2.2}$$

where η—dynamic viscosity.

Thus, having solved Eq. (2.1), and having received a value of v_φ it is possible to estimate values of stress of deformation, and consequently, values of intensity of light transmission. It allows us to receive the anisotropy parameter:

$$A \sim I' \sim \sigma_{r\varphi}, \tag{2.3}$$

where A—order parameter of mesogene system, I'—intensity of light transmission; $\sigma_{r\varphi}$—shear stress in local investigated area.

It would be useful to consider the anisotropy parameter as the certain sum of average orientation parameters of separate molecules in condition of rest and the

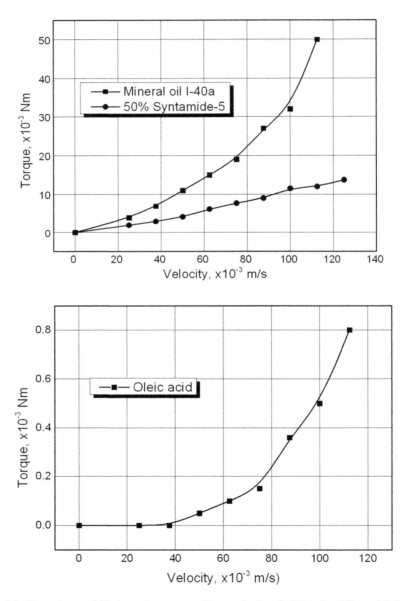

Fig. 2.1 Dependence of friction pair torque on linear velocity of sliding for different lubricants

rotary part of anisotropy depending on angular speed and radius. Though the mesogene material in an initial condition will have some initial (static) level of anisotropy, for simplicity we will consider anisotropy of the system in a condition of rest equal to zero.

The anisotropy caused by shear, which we will assume depends on components of speeds (parallel and perpendicular) double refraction and on the corner between them which, in theory, should arise at passage of a plainly polarized beam through the anisotropic environment. Unfortunately, more detailed theoretical calculations in the given assumptions present, inconveniently, only general values of the photocurrent being registered on the device so there is no possibility to analyze separately ordinary and unordinary beams arising at possible double refraction.

Experimental Results

For working off of a technique polarizing tribometry a series of experiments has been conducted. As objects for tests of the new device we used two types of fluid mesogene lubricants on oil and water base. Those were solutions of oleic acid in mineral oil and water solutions of non-ionic surfactants (see Figs. 2.1 and 2.2).

It was revealed that, for both types of modeling lubricants, the change of the photo-answer depending on shear speed and the thickness of working backlash is

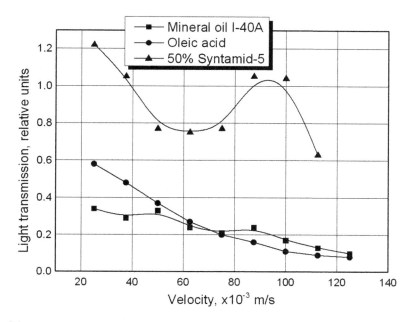

Fig. 2.2 Dependence of light transmission on linear velocity of sliding for different lubricants

indicative. Deviations from Newtonian type of flow were increased at decreasing of shear speed and at reduction of lubricating layer thickness. The evidence was received that when movement in friction pair stops, there is no fast return of the tribo-system to initial state.

Such hysteresis of orientation effect ("molecular memory" about the occurred shear) can serve as one of the additional characteristics of lubricant's "mesogenity". From tests of diluted water solutions of majority ionic surfactants, the photo-answer has not been found yet.

Conclusion

Thus, some main principles of polarizational tribometry and its instrument realization were described. The models described of thin lubricating layer rheology were proposed, the possibilities of recording in these models for the object optical anisotropy were calculated. All that was illustrated in experimental results conducted on some model mesogenic lubricants.

Acknowledgments The work was supported by the Ministry of Education and Science of the Russian Federation (project No. 9.700.2014/K).

References

1. Usoltseva NV, Godlevskiy VA (2013) Nanomaterials in tribological processes. In: Proceedings of international scientific and technical conference named after Leonardo Da Vinci. Wissenschaftliche Welt, Berlin, pp 227–234, 10–13 May 2013
2. Ermakov SF, Parkalov VP, Shardin VA, Shuldykov RA (2004) The effect of liquid crystalline additives on the tribo-engineering characteristics of dynamically contacting surfaces and the mechanism of their friction interaction. J Frict Wear 25(2):87–92
3. Kolesnikov VI, Ermakov SF, Sychev AP, Mulyarchik VV (2009) Influence of the alkyl radical on the optical activity of cholesteric liquid-crystal nanomaterials. Russian Eng Res 29(8):794–798
4. Levchenko VA (2008) Nano-tribology. The modern tribology: results and perspectives. LKI, Moscow, pp 324–325 (in Russian)
5. Oswald P, Pieranski P (2005) Nematic and cholesteric liquid crystals: concepts and physical properties illustrated by experiments. Taylor and Francis Group, London, p 618
6. Berezina EV, Godlevskiy VA, Fomichev DS, Korsakov MN (2012) Installation for investigation of optical properties of lubricants in conditions of shear. Problems of machine science: tribology—to machinery. In: Proceedings of all-Russia conference, Moscow, vol 1, pp 238–241 (in Russian)
7. Godlevskiy VA, Volkov AV (2010) Mathematical models of lubrication processes in technical tribo-systems. IvSU, Ivanovo, 140 (in Russian)

Chapter 3
Auto-Ignition Problem Titanium of Oxygen and Possible Ways of Solving

V.I. Bolobov

Abstract We note the negative feature of titanium alloys' spontaneous combustion in an oxygen environment, which prevents their widespread use in autoclave equipment operating with oxygen. We discuss the basic tenets of the theory of metals in fire destruction with an explanation of the anomalous ability of titanium to spontaneous combustion in oxygen. We concluded that the exclusion of self-ignition of titanium alloys in oxygen environment autoclaves can be achieved by developing technical measures to prevent the heating of the potential sites of friction titanium structures to auto-ignition temperature T * alloys at these partial pressures of gaseous reactants.

Keywords Titanium alloys · Oxygen gas · Ignition at destruction · Critical pressure and temperature · Juvenile surface · Adsorption

Introduction

Because of the availability and high oxidizing ability of oxygen gas, it is widely used in various fields of technology, particularly in the autoclave process for milling nonferrous and noble metals, nickel-pyrrhotite, sulphide nickel-cobalt, cobalt-arsenical concentrates, etc. [1]. To improve the speed and better oxidation reagents such processes are carried out at elevated temperatures and under the maximum possible pressure of oxygen or oxygen-air mixture (OAM). This is done to intensify the process of saturation oxygen or pulp OAM served under mechanical stirrer autoclave.

Carrying out these processes at elevated temperatures and pressures using a highly aggressive and abrasive media puts increased demands on the chemical

V.I. Bolobov (✉)
National Mineral Resources University (University of Mines), St. Petersburg, Russia
e-mail: boloboff@mail.ru

© Springer International Publishing Switzerland 2015
A. Evgrafov (ed.), *Advances in Mechanical Engineering*, Lecture Notes in Mechanical Engineering, DOI 10.1007/978-3-319-15684-2_3

resistance and the mechanical properties of structural materials used in reactor equipment, the most satisfied titanium and its alloys.

These materials are characterized by high corrosion resistance in sulfuric acid, hydrochloric acid and chlorine environments with a uniform character of corrosion of welds. Relatively high heat resistance and thermal stability, as well as good erosion resistance at elevated temperatures, allow the use of titanium and titanium alloys as the materials of the reactor equipment at temperatures up to 250 °C [1].

However, for all its positive qualities, titanium alloys have a significant negative feature, which prevents their widespread use in autoclave processes carried out using oxygen. It is a potential threat to yield titanium equipment damage due to the anomalous ability of titanium and its alloys that, under certain conditions, self-ignite in an oxygen environment.

Cases of spontaneous combustion of titanium valves and impeller mixers in oxygen have occurred, as in laboratory studies, and at the stage of pilot tests autoclaves. For example, oxidative leaching nickel-cobalt matte [1], creating a homogeneous reactor by Union Carbide Nuclear Corporation [2, 3], the autoclave low acid leaching of zinc from sphalerite concentrate at the company Hudson Bay Mining and Smelting (HBM & S) [4], as well as the development of autoclave technology in the experimental shop of MMC Norilsk . The latter cases were caused by fire violations of technological instructions for the operation of autoclave equipment, the specific reason of fire having not been established.

Large economic losses as a result of spontaneous combustion, as well as seemingly random and inexplicable phenomena that have caused it, the vast majority of non-ferrous metallurgy technology using oxygen and FAC used autoclaves made of high-chromium-nickel steels and alloys were subjected to much more severe corrosion, abrasion and cavitation compared with titanium. This significantly limits the lifetime of the equipment, reduces its performance and worsens the economic indicators.

To date, domestic work by (Nikolaeva S.A., Borisova E.A., Zashikhina T.N., Bardanov K.V., Deryabina V.I., Kolgatin N.N., Lukyanov O.P., etc.) and foreign (Littman F.E., Church F.M., Kinderman E.M., Jackson J.D., Boyd W.K., Miller P.D., etc.) scientists found that spontaneous combustion of titanium and its alloys in oxygen can occur only in the destruction of the metal, even in the surface layer, with advent of juvenile (freshly formed) metal surface. Thus a given fire alloy requires its defined minimum pressure of oxygen—a critical pressure P* fire. Since autoclaves, especially with moving parts, can not completely exclude the possibility of such a surface (e.g., as a result of friction or scratch), the threat of spontaneous combustion equipment of titanium and its alloys can be eliminated only when the process of understanding the mechanism of ignition of metals at the destruction and impact on its critical parameters of the various factors.

Below are the main tenets of the theory [5] of fire metal fracture with an explanation of the anomalous ability of titanium to spontaneous combustion in oxygen, from which we develop conclusions about the conditions of safe use of this material.

Basics of the Theory of Spontaneous Combustion of Metals

Heating of the Fragments of Destruction as a Key Factor Explaining the Spontaneous Combustion of Materials

In establishing a possible mechanism of spontaneous combustion, we are guided by known fact [6, 7] that the plastic deformation of metals, regardless of the impact, the vast majority of work (more than 90 %) is converted into heat.

Given that failure is the final stage of plastic deformation, it is believed that the vast majority of A_p expended in the destruction of the metal is converted into heat Q, which is released in the volume and destroys the part (or all) consumed in heating fragments of destruction. For this reason, the destruction of fire exposed fragments, already warmed up to the moment of interaction with oxygen to a temperature T^*, which is the sum of the initial temperature T_0 and heating ΔT due to the heat released in the amount of destructible, is

$$T^* = T_0 + \Delta T \tag{3.1}$$

In this case, the temperature T^* is the critical parameter characterizing the propensity to spontaneous combustion of the metal associated with the value of another critical parameter—ignition pressure P^*.

Analyzed heating ΔT fragments fracture of titanium alloys breaks tensile specimens. The quantity of heat Q, released during the destruction, was compared to the work of destruction A_p,

$$Q = k\,A_p, \tag{3.2}$$

where k—describes the share of A_p passed into heat.

Work A_p is expressed as the product of the breaking load P on the path dL, during which it is performed. When submitting blasted volume as $V = F \cdot dLV = F{\cdot}dL$ (where F—cross-sectional area of the sample in the gap) it is the specific work of fracture

$$A_p/V = P \cdot dL/F \cdot dL = P/F = S_k, \tag{3.3}$$

where S_k—the true fracture stress.

We assume that the destruction of the sample at the final stage takes place almost instantaneously and fracture fragments with difficult heatsink exposed by the extracted heat Q adiabatic heating, the value of which in view of (2.3) can be calculated by the formula

$$\Delta T \cong kS_k/\rho \cdot \bar{c}_p, \tag{3.4}$$

where $\rho, \bar{c}_p$—density and the average value of the specific heat-blasted metal at temperatures $T_0 - T^*$.

Table 3.1 Calculated values $S_k, \Delta T, T^*$ titanium α-alloys in comparison with the experimentally established values of the critical pressure P^* fire alloys in oxygen at various test temperatures T_0

Alloy grade	$T_0(K)$	P*(MPa)	σ_B(MPa)	ψ	S_k(MPa)	$\Delta T(K)$	$T^*(K)$
VT1-0	473	2.9	250	0.73	576	209	682
VT1-0	293	2.3	470	0.73	1,083	405	698
Ti-15Zr	293	1.6	540	0.65	1,155	449	742
OT4-1	293	1.5	680	0.44	1,160	442	735
PT17	293	1.1	925	0.26	1,235	455	748
PT3V	293	0.9	770	0.41	1,266	470	763
OT4-1	473	0.7	450	0.56	879	297	770

Values S_k were calculated according to the values in the σ_B (apparent yield strength) and ψ (relative reduction) analyzed alloys.

$$S_k = \sigma_B \times (0.8 + 2.06 \times \psi) \qquad (3.5)$$

The calculated data were compared with the results of experiments conducted by the author under uniaxial tensile rupture of cylindrical samples of various titanium alloys in gaseous oxygen (Table 3.1).

The results of the calculation were that the fragments of destruction of all analyzed samples of alloys are capable of undergoing significant self-heating (ΔT up to 470 K), the magnitude of which is individual for each alloy. These circumstances explain the marked influence of the literature on the value of P^* due to a variety of factors, in particular, the alloy composition and the rate of appearance of juvenile surface.

Thus, the table shows that with increasing temperature T^* (from 682 to 770 K), to which the self-igniting fragments are capable of destruction alloys, oxygen pressure P^*, necessary for their fire, are monotonically decreasing (from 2.9 to 0.7 MPa). Constant temperature tests ($T_0 = 293$ K) in a similar way to the P^* value correlates with heating samples: with increasing ΔT from 405 to 470 K pressure, fire alloys decrease from 2.3 to 0.9 MPa. Because the value of ΔT materials is dictated by the failure of the sample, and, in turn, (4) that the value of the failure stress, the results indicate the existence of a table dependence of the critical pressure P^* fire on the strength properties of the alloy: the higher the voltage required to break the specimen, the {AU: the "higher" or the "lower" the oxygen pressure?} the oxygen pressure at which the fracture fragments ignited. For this reason, titanium iodide, characterized by the lowest strength of titanium alloys (S_k 550 MPa at $T_0 = 293$ K) and, as a consequence of undergoing the destruction of the smallest heating up, different high resistance to spontaneous combustion (P = 7.5 MPa [8]). With the introduction of the titanium alloying elements that enhance the strength of the material, the quantity ΔT fracture fragments increases, which leads to a reduction in pressure of oxygen at which these fragments ignite.

On the other hand in the calculations ΔT it is assumed that the final phase of sample failure occurs almost instantaneously, resulting fragments blasted volume capable of undergoing adiabatic heating. Naturally, the low rate of appearance of juvenile surface that holds tensile specimens without breaking, these conditions are not met and the heat released during plastic deformation is dissipated into the environment. For this reason, the plastic deformation of the titanium alloy specimens during tensile stresses are less destructive, although the accompanied juvenile form surface does not lead to ignition of materials at very substantial pressures of oxygen (e.g., an alloy VT5-1 under PO2 at 10 MPa [8]). Inadequate rate of appearance of fragments with juvenile surface characteristic when tested titanium plates in high-pressure oxygen bending metal when fire is observed only in the case of formation of fracture [9].

Quantitative Relationship Between the Critical Parameters T* and P* in Fire

Experimentally established fact that the influence of the pressure of oxygen on the propensity to spontaneous combustion of metals indicates the author's opinion, the fact that the limiting stage of interaction in this case is the chemical adsorption of oxygen on the surface of the formed fragments juvenile destruction. The rate of reaction is proportional to oxygen for a diatomic gas, $\sqrt{P}$ and for different temperatures is described by the equation

$$\frac{\partial m}{\partial t} = K_0 \left(P/P_{0,1}\right)^{0,5} \cdot \exp(-E/RT), \ kgO_2/\left(m^2 \cdot s\right), \tag{3.6}$$

where K_0, E—activation energy and pre-exponential factor of the Arrhenius equation for the rate of adsorption of O_2 juvenile metal surface at an oxygen pressure $P_{0,1} = 0.1$ MPa, and the fire conditions fragments determined by the critical conditions for thermal explosion theory Semenov-Frank-Kamenetsky [10] for the heterogeneous reaction.

$$\delta_{cr} = \frac{Q}{\alpha} \cdot \frac{K_o \cdot E}{R \cdot (T^*)^2} (\bar{P})^{0.5} \cdot \exp(-E/RT^*) = \frac{1}{e}, \tag{3.7}$$

where Q—specific heat effect of the interaction of oxygen with the metal on the surface of the juvenile (assumed equal to the heat of formation of the corresponding oxide); α—the total heat transfer coefficient of the destruction of the fragment; $\bar{P} = P * /P_{0,1}$.

Verification of compliance parameters T^*, P^* materials criterion Eq. (3.7) was performed using the experimental determinations of the author of the $P^* = f(T^*)$ for titanium alloys tensile specimens to rupture (Table 3.1), as well as iron and stainless steel with rupture pressure of the oxygen tubes (results presented in Fig. 3.1). To obtain the true temperature (T^*) to the values of temperature T_0 of Fig. 3.1 is added on the corresponding calculated values of heating ΔT, accounting for up to 15 % of

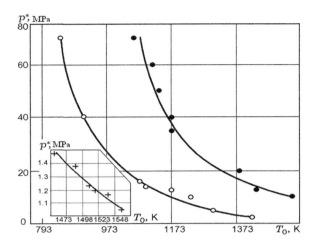

Fig. 3.1 Critical pressure P* ignition materials as a function of test temperature T_0: technical iron (°), steel 12Cr18Ni10Ti (•)—the destruction of oxygen pressure tubes; technical iron (+) gap cylindrical specimens under uniaxial tension in the decomposition products N_2O [11]

T_0. In the calculations it is assumed that all of the studied titanium α-alloys kinetic parameters of the rate equations of their interaction with the surface of juvenile oxygen close to each other, and α = const (P).

After taking the logarithm and grouping variables, Eq. (3.7) has the form

$$0.5\,\overline{\ln P} - 2\ln T^* - B/T^* + A = 0, \tag{3.8}$$

where $B = E/R$; $A = \ln(Q \times K_0 e \times E/R/a)$, from which one could conclude that the dependence of $0.5\overline{\ln P} - 2\ln T^* = f(1/T^*)$, built on the values of P*, T* of Table 3.1 and Fig. 3.1 should represent a straight line with a slope B, which, as follows from Fig. 3.2 is indeed the case. This may serve as a confirmation that the conditions of fire metal fracture indeed characterized by two critical parameters T*, P* and described by the equation of thermal explosion.

Nature and Conditions of Heat Exchange Initiators Fire

Based on the results of fractographic observations fracture of titanium alloys and iron technical, it was concluded that the primary initiators of spontaneous combustion of materials are tiny destruction of the order of the grain size of the metal (5–20 mkm), are in the most difficult, the heat exchange with the bulk of the sample (Fig. 3.3) {AU: The previous sentence is indecipherable}. Thus their ignition as the tensile specimens to fracture and breaking the tubes at pressure medium takes place

0.5·ln $\bar{P}$ - 2·ln T⁺

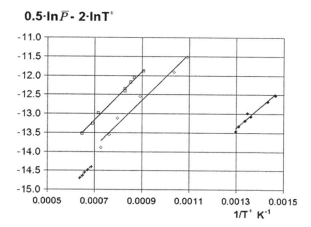

Fig. 3.2 Approximation calculations and experimental values of the critical parameters P*, T* dependence criterion equations of the theory of thermal explosion: +— technical iron in O₂-containing in mixture; ◇ - technical iron, □ - stainless steel 12Cr18Ni10Ti, ◆—Titanium α—alloys in O₂

Fig. 3.3 Mikrofraktogramma halves sample surface alloy OT4-1 after his break in oxygen at various pressures (*—extinct hearth fire): a—P = 1.4 MPa, X 24; b—P = 1.6 MPa, X 24; c—P = 2.1 MPa, X 72

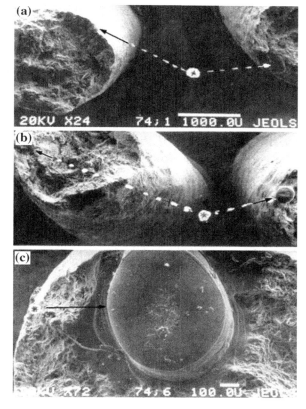

Fig. 3.4 Type metal grains
on the surface of the crack
fracture of the tubular
stainless steel sample
(T = 1,373 K, P_{O2} = 20 MPa
(P = P*) X 600)

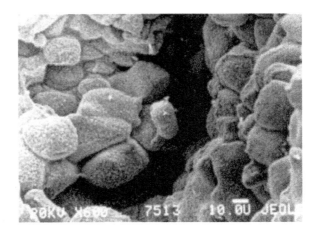

in the same heat transfer conditions: when the pieces are still in a narrow crack
stagnant fracture filled with a working pressure of oxygen (Fig. 3.4).

For this reason, the total heat transfer coefficient values α for both analytes
methods fracture was calculated from the formula (3.9) for the narrow gap stagnant

$$\alpha \cong \alpha_\alpha \cong \lambda_g p / X \tag{3.9}$$

where X—the average cell size of the fracture surface mounted fractometrical way
(~ 10 microns for samples of titanium alloys, ~ 20 microns for iron and steel tubes
for samples ~ 70 microns iron), λ_g^P—the thermal conductivity of oxygen at P*, T*.

Kinetic Parameters of the Rate Equation of Oxygen Adsorption on the Surface of Juvenile Metallic Structural Materials

Equation (3.4) after substituting values of P*, T*, α, Q solved by the method of least
squares, resulting in the calculated parameters of the Arrhenius equation for the rate
of oxygen adsorption on the surface of metallic materials juvenile at P_{O2} = 0.1 MPa:

E = 56.4 $\pm$ 9,2 kJ/mol (13.5 $\pm$ 2.2 kcal/mol), K_0 = 0.72 $\pm$ 0.1 $kg_{O2}/m^2/s$ and
E = 50.94 $\pm$ 10.4 kJ/mol (12.2 $\pm$ 2.5 kcal/mol), K_0 = 1.17 $\pm$ 0.2 $kg_{O2}/m^2/s$—for iron
in an oxygen mixture and pure oxygen, respectively;

E = 52.9 $\pm$ 4.4 kJ/mol (12.6 $\pm$ 1.1 kcal/mol), K_0 = 0.54 $\pm$ 0.04 $kg_{O2}/m^2/s$—for
Cr-Ni steel in O_2;

E = 44.5 $\pm$ 10.5 kJ/mol (10.6 $\pm$ 2.5 kcal/mol), K_0 = 8.3 $\pm$ 1.6 $kg_{O2}/m^2/s$—for
titanium α-alloys in O_2.

Regardless of the method of destruction of samples obtained values of the
activation energy for all materials are close to each other and have order

characteristic of the process of oxygen chemisorption on metals (11.5 kcal/mol—for Ta, 14,2-for Nb, 17,6-to V [12]) and their oxides (13.8 kcal/mol for Fe_2O_3 [13]) that supports the hypothesis that the rate-limiting step in the process of adsorption of oxygen in the destruction of metal ignition.

Comparing the values of the kinetic parameters it can be seen that the value of pre-exponential factor K_0 for titanium alloys are significantly higher and the activation energy E is lower than the rest of the materials studied, which provides for the same temperature significantly greater rate of reaction of oxygen with the surface of the titanium juvenile. For example, when $T = T*_{VT1-0} = 682$ K values $\partial m/\partial t$ of titanium and iron at $P_{O2} = 0.1$ MPa respectively 1, 2 10^{-3} и 5,5 10^{-5} $kg_{O2}/m^2/s$. This circumstance, as well as abnormally high solubility of oxygen in titanium (14.5 % by weight compared with 10^{-1}–10^{-3} % for other metals) and the accompanying thermal effect (6 times the value of Q for iron), explain the unique ability of titanium and its alloys in a fire in the destruction in an oxygen atmosphere.

Solutions to the Problem

According to the proposed theory the possibility of fire titanium alloys (and other metal materials) in oxygen at the destruction determined by the achievement of the critical conditions under which each value of the oxygen pressure P corresponds to its temperature T^* fracture fragments (regardless of whether thereby achieved T^*) and vice versa. In this regard, in order to prevent ignition of titanium constructions in terms of friction necessary to eliminate the situation of the friction pieces of heating to a temperature T^* corresponding to the given partial pressure of oxygen.

With regard to establishing the critical ignition temperature T^* titanium alloys, its values or that the oxygen partial pressure P_{O2} in the mixture can be calculated from the formula (3.10) from [14]

$$\frac{1,44 \cdot 10^{-9} \cdot (T^*)^2}{\exp\left(-\frac{5300}{T^*}\right)} = \frac{\left(P^*_{O_2}\right)^{0,5}}{1 + 0,52 \cdot P^{0,5}_{N_2,H_2O}}, \qquad (3.10)$$

where T^*—temperature in К; $P^*_{O_2}$, P_{N_2,H_2O},—partial pressures of the reactants in MPa.

Thus, the limiting value of the partial pressures of oxygen in the oxygen-air mixtures autoclave leaching processes non-ferrous metals (1.0 MPa at limiting the content of O_2 in a mixture of 23 %) in accordance with this formula corresponds to the critical temperature T* juvenile surface equal 520 °C. Accordingly, up to this temperature and unacceptable heating of the potential sites of titanium friction elements.

It must be emphasized that the values of the coefficients in Eq. (3.10), fitted in [7, 14] from the values of the critical parameters T^*, P^* arbitrarily taken various titanium alloys and are thus common to all of them. For this reason, it can be assumed that this equation is valid for all alloys of titanium, regardless of the brand.

Thus, we can conclude that the exclusion of self-ignition of titanium alloys in oxygen in an autoclave with mixing devices may be provided by the development of technical measures to prevent the heating of the potential sites of friction titanium structures to auto-ignition temperature T^* alloys at these partial pressures of gaseous reactants.

References

1. Naboichenko SS, Ni LP, Schneerson YM, Chugaev LV (2002) Autoclave hydrometallurgy of nonferrous metals. In: Corresponding Member of Naboichenko RASS (ed) Ekaterinburg: USTU, 940 pp (in Russian)
2. Littman FE, Church FM, Kinderman EM (1961) A study of metal ignitions. The spontaneous ignition of titanium. J Less-Common Metals 3:367–378 (in Russian)
3. Littman FE, Church FM, Kinderman EM (1961) A study of metal ignitions. The spontaneous ignition of zirconium. J Less-Common Metals 3:378–397
4. Barth TR, Hair ATC, Meier TP (1998) Zinc and lead processing the metallurgical society of CIM (Hudson Bay Mining and Smelting Company, Ltd PO Box 1500, Flin Flon, Manitoba R8A 1N9)
5. Nadai A (1969) Plasticity and fracture of solids. Wiley, New York
6. Larikov LN, Yurchenko YuF (1985) The structure and properties of metals and alloys. In: Handbook. Thermal properties of metals and alloys. Kiev. Navukova Dumka (in Russian)
7. Bolobov VI, Podlevskikh NA (2007) Theory of fire metal fracture. Phys Burning Explos 43(4):39–48 (in Russian)
8. Borisova EA, Bardanov KV (1963) Fire titanium alloys in oxygen-containing media. Metall Heat Treat Met 2:37–40 (in Russian)
9. Deryabina NN, Kolgatin NN, Lukyanov OP, etc. (1971) Ignition and other low-alloy titanium α-alloy at break in oxygen gases. Phys Chem Mech Mater 1:16–19 (in Russian)
10. Frank-Kamenetsky DA (1987) Diffusion and heat transfer in chemical kinetics. Nauka, Moscow (in Russian)
11. Bolobov VI, Makarov KM, Steinberg AS, Drozhzhin PF (1992) About ignited compact samples when a juvenile metal surface. Phys Burning Explos 28(5):8–11 (in Russian)
12. Fromm E, Gebhardt E (1980) Gases and carbon in metals. Moscow, Metallurgy (in Russian)
13. Benard J, Telbot J (1948) Sur la cinetique de la reaction d'oxydation du fer dans sa phase initiale. Compt Rend 226(11):912–914
14. Bolobov VI (2011) On the calculation of the critical pressure fire titanium alloys vapor gas mixtures autoclaves. Non-ferrous Metals, 10:94–97 (in Russian)

Chapter 4
Control of Biped Walking Robot Using Equations of the Inverted Pendulum

Valeriy Tereshin and Anastasiya Borina

Abstract This paper presents a control method for a 3-D biped robot walking in different modes: walking in the up or down direction, walking up stairs or down stairs, and rotation. The control strategy is developed on a simplified model of the robot and then verified on a more realistic model (Pott A, Kecskeméthy A, Hiller M (2007) Mech Mach Theory 11:1445–1461). This paper includes the results of the numerical solutions, showing stepping on and off a 0.1 m high platform, walking in the down direction (surface slope angle is 0.02 rad), and rotation on the second step by a 0.3 rad angle.

Keywords Walking robot · Movement control · Dynamic stability

Introduction

Today making walking robots is very important and attracts an interest in industry and in academia [6, 12]. In 2000 Sony Corporation announced the development of a small biped walking robot "SDR-3X". The robot can perform basic movements such as walking and changing direction, balancing on one leg, kicking a ball and dancing. In 2001 Honda engineers created ASIMO with 34 Degrees of Freedom that help it walk and perform tasks much like a human. Big Dog developed by Boston Dynamics in 2005 is a rough-terrain robot that walks, runs, climbs and carries heavy loads.

Legged robots have high mobilities due to the similar morphologies with the humans' lower limbs [9]. So they can travel wherever humans can. Most of the legged robots are statically stable, but they are very heavy and slow [2, 13]. Dynamically stable walking robots have small feet, they are more mobile and light [1]. During the

V. Tereshin (✉) · A. Borina
Saint Petersburg State Polytechnic University, Saint Petersburg, Russia
e-mail: terva@mail.ru

A. Borina
e-mail: kamchatka1@rambler.ru

© Springer International Publishing Switzerland 2015
A. Evgrafov (ed.), *Advances in Mechanical Engineering*, Lecture Notes in Mechanical Engineering, DOI 10.1007/978-3-319-15684-2_4

walking robots lean only by one foot on the ground for an appreciable period of time [3]. For a biped robot, the support area is small. Because it is passively unstable and non-linear it is not easy to design a walking controller. The control system should provide processing of information about the area [7, 11], making decisions about the movement, and control over the implementation.

Model of Potential Containers

Figure 4.1 shows the structure of the biped and coordinates used to describe the configuration of the system.

Simulation of the Center of Gravity Projection Trajectory During Walking on Stairs and in the Down Direction

In [4, 5] we have obtained the equations of spatial movement, investigated the possibility of automatic control for stable and constant biped walking, and defined the desired time and place for touchdown at the end of the step and at the beginning of the next one.

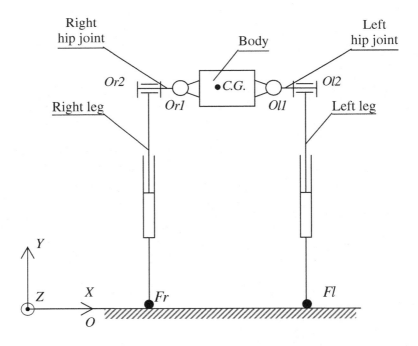

Fig. 4.1 Kinematic model of a biped robot

This paper proposes a method of biped walking control in different modes: walking in the up or down direction, walking up stairs or down stairs, and rotation.

Walking up stairs differs from the walking in the up or down direction by the height of setting foot y_l.

Method of biped walking control is presented in [5]. Let the width of the step be constant $z_l = 0.25$ m. The time for touchdown is determined from the condition of equality of the longitudinal speed to its initial value. Figures 4.2 and 4.3 present the results of the numerical solutions of the system of equations, presented in [4], under the given conditions and parameters of the walking robot in the case of walking up stairs $(y_l = 0.1$ m$)$ and down stairs $(y_l = -0.1$ m$)$.

Figure 4.3 shows that the walking up stairs performed in the second step, and down stairs on the fifth. In general, the robot makes seven steps.

We consider a case of constant walking in the down direction. Let $y_l = 0, -0.1, -0.02, \ldots, -0.06$ m. The numerical solutions of the system of equations, presented in [4], are given in Figs. 4.4 and 4.5.

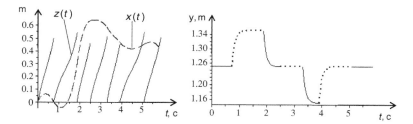

Fig. 4.2 **a** The longitudinal and transverse coordinates and **b** the height of COG during walking on stairs

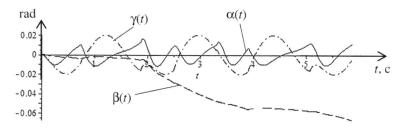

Fig. 4.3 Finite rotation angles or the robot body during walking on stairs

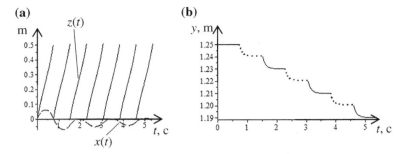

Fig. 4.4 **a** The longitudinal and transverse coordinates and **b** the height of COG during walking in the down direction

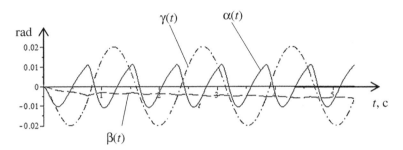

Fig. 4.5 The finite rotation angles or the robot body during walking in the down direction

Simulation of the Center of Gravity Projection Trajectory During the Rotation

We describe an example of the turn angle α_y to the left (see Fig. 4.6).

We define longitudinal $z_i(T)$ and cross- $x_i(T)$ the coordinates of the center of gravity of the robot body at the time of the end of the current step T in the new rotated $(i + 1)$th coordinate system. The transition matrix to the $(i + 1)$th system of the ith coordinate system can be represented as

$$H_{i+1,i} = \begin{bmatrix} cos\,(\alpha_y) & 0 & -sin\,(\alpha_y) & (z_i(T) + x_i(T)tg(\alpha_y))sin\,(\alpha_y) \\ 0 & 1 & 0 & 0 \\ sin\,(\alpha_y) & 0 & cos\,(\alpha_y) & -(z_i(T) + x_i(T)tg(\alpha_y))cos\,(\alpha_y) \\ 0 & 0 & 0 & 1 \end{bmatrix} \quad (4.1)$$

Fig. 4.6 The rotation of the system $X_iY_iZ_i$ by the angle α_y

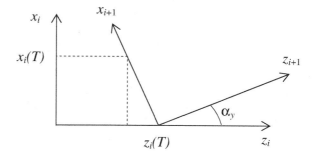

At time T, the center of gravity of the robot body in the system i has coordinates

$$R_i = \begin{bmatrix} x_i(T) \\ y_i(T) \\ z_i(T) \\ 1 \end{bmatrix}. \tag{4.2}$$

We define these coordinates in the system $(i + 1)$

$$R_{i+1} = H_{i+1,i} \cdot R_i$$

$$= \begin{bmatrix} \cos(\alpha_y) & 0 & -\sin(\alpha_y) & (z_i(T) + x_i(T)tg(\alpha_y))\sin(\alpha_y) \\ 0 & 1 & 0 & 0 \\ \sin(\alpha_y) & 0 & \cos(\alpha_y) & -(z_i(T) + x_i(T)tg(\alpha_y))\cos(\alpha_y) \\ 0 & 0 & 0 & 1 \end{bmatrix} \cdot \begin{bmatrix} x_i(T) \\ y_i(T) \\ z_i(T) \\ 1 \end{bmatrix} =$$

$$= \begin{bmatrix} x_i(T)\cos(\alpha_y) - z_i(T)\sin(\alpha_y) + z_i(T)\sin(\alpha_y) + x_i(T)\tan(\alpha_y)\sin(\alpha_y) \\ y_i(T) \\ x_i(T)\sin(\alpha_y) + z_i(T)\cos(\alpha_y) - z_i(T)\cos(\alpha_y) - x_i(T)\tan(\alpha_y)\cos(\alpha_y) \\ 1 \end{bmatrix}$$

$$= \begin{bmatrix} x_i(T)\cos(\alpha_y) + x_i(T)\frac{\sin^2(\alpha_y)}{\cos(\alpha_y)} \\ y_i(T) \\ x_i(T)\sin(\alpha_y) - x_i(T)\frac{\sin(\alpha_y)}{\cos(\alpha_y)}\cos(\alpha_y) \\ 1 \end{bmatrix} = \begin{bmatrix} \frac{x_i(T)}{\cos(\alpha_y)} \\ y_i(T) \\ 0 \\ 1 \end{bmatrix} \tag{4.3}$$

Therefore, in the system $(i + 1)$ transverse and longitudinal coordinates of the center of gravity can be represented as

$$x_{i+1} = \frac{x_i(T)}{\cos(\alpha_y)}, \qquad z_{i+1} = 0. \tag{4.4}$$

Similarly, we write the expression for the velocity of the center of gravity in the system $(i + 1)$,

$$V_{i+1} = A_{i+1,i} \cdot V_i$$

$$= \begin{bmatrix} \cos(\alpha_y) & 0 & -\sin(\alpha_y) \\ 0 & 1 & 0 \\ \sin(\alpha_y) & 0 & \cos(\alpha_y) \end{bmatrix} \cdot \begin{bmatrix} \dot{x}_i(T) \\ \dot{y}_i(T) \\ \dot{z}_i(T) \end{bmatrix} = \begin{bmatrix} \dot{x}_i(T)\cos(\alpha_y) - \dot{z}_i(T)\sin(\alpha_y) \\ \dot{y}_i(T) \\ \dot{x}_i(T)\sin(\alpha_y) + \dot{z}_i(T)\cos(\alpha_y) \end{bmatrix},$$

$$\tag{4.5}$$

where $A_{i+1,i}$—cosine matrix, V_i—column-vector of the speeds at time T onto the i-axis. Therefore, we obtain the projection of the transverse and longitudinal velocities in the rotated angle α_y system $i + 1$ as

$$\dot{x}_{i+1} = \dot{x}_i(T)\cos(\alpha_y) - \dot{z}_i(T)\sin(\alpha_y), \\ \dot{z}_{i+1} = \dot{x}_i(T)\sin(\alpha_y) + \dot{z}_i(T)\cos(\alpha_y). \tag{4.6}$$

So, rotation control is performed according to the formulas obtained in [5], adjusted for the rotation:

$$x_l = \frac{x_i(T) \cdot \delta}{2\cos(\alpha_y)} + \frac{1}{k}(\dot{x}_i(T)\cos(\alpha_y) - \dot{z}_i(T)\sin(\alpha_y)) \tag{4.7}$$

where δ—the coefficient of stability in the transverse direction, T—the end time of the current step; $k = \sqrt{g/L}$; L—height of center of gravity; g—the acceleration of free fall. We define coordinates of center of gravity of the robot body in the ith system, and let it coincide with the global [8]. The transition matrix in the global system can be represented as

$$H_{i,i+1} = \begin{bmatrix} \cos(\alpha_y) & 0 & \sin(\alpha_y) & 0 \\ 0 & 1 & 0 & 0 \\ -\sin(\alpha_y) & 0 & \cos(\alpha_y) & z_i(T) + x_i(T)tg(\alpha_y) \\ 0 & 0 & 0 & 1 \end{bmatrix} \tag{4.8}$$

Column-vector coordinates of the center of gravity in the global system can be represented as

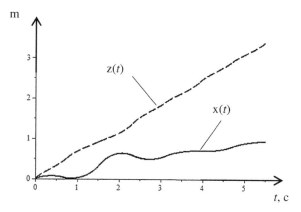

Fig. 4.7 The longitudinal and transverse coordinates during *left* rotation on the second step in the global coordinate system

$$R_i = H_{i,i+1} \cdot R_{i+1}$$

$$= \begin{bmatrix} \cos(\alpha_y) & 0 & \sin(\alpha_y) & 0 \\ 0 & 1 & 0 & 0 \\ -\sin(\alpha_y) & 0 & \cos(\alpha_y) & z_i(T) + x_i(T)tg(\alpha_y) \\ 0 & 0 & 0 & 1 \end{bmatrix} \cdot \begin{bmatrix} x_{i+1} \\ y_{i+1} \\ z_{i+1} \\ 1 \end{bmatrix} =$$

$$= \begin{bmatrix} x_{i+1}\cos(\alpha_y) + z_{i+1}\sin(\alpha_y) \\ y_{i+1} \\ -x_{i+1}\sin(\alpha_y) + z_{i+1}\cos(\alpha_y) + z_i(T) + x_i\tan(\alpha_y) \\ 1 \end{bmatrix}. \qquad (4.9)$$

Therefore

$$x_i = x_{i+1}\cos(\alpha_y) + z_{i+1}(T)\sin(\alpha_y),$$
$$z_i = -x_{i+1}\sin(\alpha_y) + z_{i+1}\cos(\alpha_y) + z_i(T) + x_i\tan(\alpha_y). \qquad (4.10)$$

Figures 4.7, 4.8 and 4.9 present the results of the numerical solutions of the system of equations presented in [4], under the given conditions and parameters of the walking robot during left rotation on the second step by the angle $\alpha_y = 0.3$.

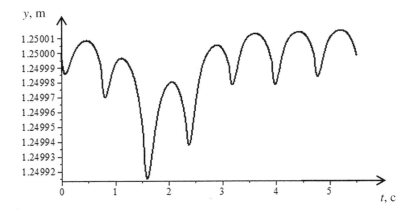

Fig. 4.8 The height of center of gravity during *left* rotation

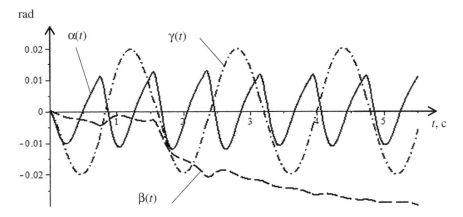

Fig. 4.9 The finite rotation angles or the robot body during *left* rotation

Conclusion

Coordinate z discontinuity is determined by the translation of the origin of the coordinate at the end of each step to the point $x = 0$, $y = 0$, $z = z(T)$. The height of the center of gravity stabilizes due to changing the lengths of the legs, and according to the calculations it is obtained almost constant.

Due to the smallness of the obtained angles α, β and γ, let them be the finite rotation angles around the axes x, y and z respectively. The angle β in this model is not regulated and its slowness was achieved by initial conditions. The angles α and γ are stabilized by the driving moments applied to the axes O2 and O1, respectively. The angle β is stabilized when the robot leans by both feet.

References

1. Andrienko P, Kozlikin D, Hisamov A, Hlebosolov I (2013) About the using of non-contact measurement systems in centrifugal stands. Sovremennoe mashinostroenie. Nauka i obrazovanie 3:623–630 (in Russian)
2. Beletsky V (1975) Dynamics of bipedal walking. Izv. AN SSSR, MTT 3:3–14 (in Russian)
3. Bertrand S, Bruneau O, Ouezdou FB, Alfayad S (2012) Closed-form solutions of inverse kinematic models for the control of a biped robot with 8 active degrees of freedom per leg. Mech Mach Theory 3:117–140
4. Borina A, Tereshin V (2013) The solution of the problem of the spatial movement of statically unstable walking robot. Sovremennoe mashinostroenie. Nauks i obrazovanie: materialy nauchno-prakticheskoi konferensii s mezhunarodnym uchastiem 20–21 iunya 2013 goda, Sankt-Peterburg. SPb.: Izd-vo Politehn. un-ta, pp 631–641 (in Russian)
5. Borina A, Tereshin V (2014) Stable and constant biped walking of humanoid robot. Sovremennoe mashinostroenie. Nauks i obrazovanie: materialy nauchno-prakticheskoi konferensii s mezhunarodnym uchastiem. Institut metallurgii, mashinostroeniya I transporta SPbGPU CH.1. SPb.: Izd-vo Politehn. un-ta, pp 160–162 (in Russian)
6. Borina A, Tereshin V (2012) The spatial movement of the walking device under the action of the control actions from the side of the legs. Sovremennoe mashinostroenie. Nauks i obrazovanie 2:177–188 (in Russian)
7. Nicola S, Vincenzo P-C (2011) A sequentially-defined stiffness model of the knee. Mech Mach Theory 46:1920–1928
8. Parenti-Castelli V, Venanzi S (2005) Clearance influence analysis on mechanisms. Mech Mach Theory 12:1316–1329
9. Petrov G (2009) Computer simulation of mechanical systems in "Model Vision". Nauchno-tekhnicheskie vedomosti SPbGPU. Informatika i Telekomunikatsii. Upravlenie 6(91):239–244 (in Russian)
10. Pott A, Kecskeméthy A, Hiller M (2007) A simplified force-based method for the linearization and sensitivity analysis of complex manipulation systems. Mech Mach Theory 11:1445–1461
11. Spong M, Holm J, Lee D (2007) Passivity-based control of bipedal locomotion. IEEE Robot Autom Mag 14(2):30–40
12. Tomomichi S (2010) Simulated regulator to synthesize ZMP manipulation and foot location for autonomous control of biped robots. Climb Walk Robot: 201–211
13. Westervelt E, Grizzle J, Chevallereau C, Choi J, Morris B (2007) Feedback control of dynamic bipedal robot locomotion. CRC Press, Boca Raton

Chapter 5
An Efficient Model for the Orthogonal Packing Problem

Vladislav A. Chekanin and Alexander V. Chekanin

Abstract The orthogonal packing problem is considered in the article. A new model for management of free spaces of containers is presented. The efficiency of the proposed model of potential containers is demonstrated on the three-dimensional orthogonal bin packing problem.

Keywords Packing problem · Orthogonal packing · Bin packing · Packing model · Recourse allocation · Optimization

Introduction

The orthogonal packing problem is an NP-completed problem [11] that deals with the optimal packing of a given set of orthogonal objects into a set of orthogonal containers under conditions of correct placement [10, 16]:

- all edges of objects packed into a container are parallel to edges of the container;
- no packed object overlaps another;
- all objects packed into a container are within the bounds of the container.

Solution of a large number of different practical optimization problems comes down to the orthogonal packing problem [9, 12, 13]. In particular this problem is actual in solving of various optimization problems in logistics, calendar planning problems, waste minimization problems in cutting and many others in industry [1, 3, 4, 8, 14].

V.A. Chekanin (✉) · A.V. Chekanin
Moscow State University of Technology "STANKIN", Moscow, Russia
e-mail: vladchekanin@rambler.ru

A.V. Chekanin
e-mail: avchekanin@rambler.ru

© Springer International Publishing Switzerland 2015
A. Evgrafov (ed.), *Advances in Mechanical Engineering*, Lecture Notes in Mechanical Engineering, DOI 10.1007/978-3-319-15684-2_5

The current state of a container packed with objects is described by a packing representation model. This model is used for management of free spaces of a container. In practice the following models are usually used: the block model [15], the node model and the virtual object model, proposed early by the authors of this paper, that provides the fastest placement among all these packing representation models [2, 5, 7]. The major disadvantage of the virtual object model is a risk of formation of local residual free spaces of containers, which are not supervised by this model and reduce density of a resulting placement.

Model of Potential Containers

The developed model of potential containers is an extension of the virtual object model. Under the potential container, placed in a container at some of its points, is understood an orthogonal object with the largest dimensions that can be placed at this point with no overlap with all packed into the container objects and with the edges of the container. The model of potential containers unlike the virtual object model describes all existing free spaces of a container, which eliminates the probability of uncontrolled local spaces of the container.

On Fig. 5.1 are shown all potential containers that can be generated in a three-dimensional container when a new object is packed into a free space of the container.

The algorithm of placement of a given sequence of objects for the model of potential containers shows a flowchart on Fig. 5.2.

Computational Experiment

The effectiveness of the model of potential containers in comparison with the virtual object model was investigated by solving the orthogonal packing problems. The carried out computational experiments on standard two-dimensional packing problems showed that the model of potential containers provides a higher density of placement than the virtual object model in less time for all tested classes of the problems. All details of this experiment are given in paper [6].

The proposed model also was investigated on the three-dimensional orthogonal packing problem. The considered test instance includes 500 three-dimensional orthogonal objects of six different types that have to be packed into an orthogonal container with dimensions $300 \times 100 \times 100$. All parameters of this orthogonal packing problem are shown in Table 5.1.

When all objects are packed into one container as an indication of quality of the resulting packing is usually used the relative density of placement S, which generally is calculated as the ratio of the total volume of the objects placed into the container to the volume V^* of an orthogonal D-dimensional container that circumscribes all packed objects:

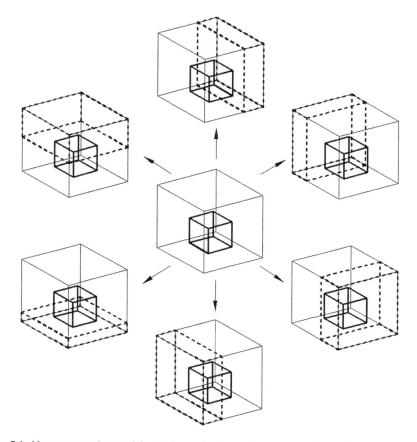

Fig. 5.1 New generated potential containers of a three-dimensional orthogonal container

$$S = \frac{\sum\limits_{i=1}^{n} \prod\limits_{d=1}^{D} w_i^d}{V^*},$$

where n—total number of packed into the container objects; w_i^d—dimension of the object i that measured along the axis d of the container; D—the dimension of the problem (in three-dimensional case $D = 3$). For the optimal (ideal) placement the relative density of placement $S = 1$.

The results of the carried out computational experiment (Table 5.2) demonstrate that the developed model of potential containers provides obtaining of more dense placement compared to the effective virtual object model.

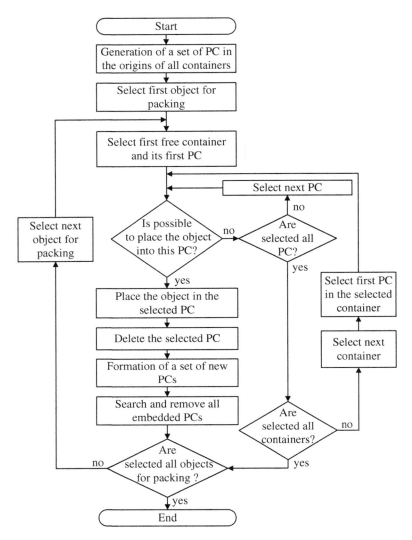

Fig. 5.2 Algorithm of placement for the model of potential containers (PC)

Table 5.1 Parameters of the test orthogonal packing problem

No.	Object dimensions	Amount of objects
1	50 × 30 × 10	20
2	10 × 30 × 60	20
3	25 × 30 × 25	20
4	25 × 25 × 25	20
5	10 × 10 × 10	20
6	5 × 5 × 5	400

Table 5.2 Result of computational experiment

No.	Packing representation model	Relative density of placement S
1	Virtual object model	0.727
2	Model of potential containers	0.810

Conclusion

A new model for management of free spaces of containers has been proposed. The developed model of potential containers describes all existing free spaces of orthogonal containers. The carried out computational experiment on the three-dimensional packing problem demonstrated the high effectiveness of the proposed model and its applicability in solving of orthogonal packing problems that are very actual in different applications in industry.

References

1. Bortfeldt A, Wascher G (2013) Constraints in container loading—a state-of-the-art review. Eur J Oper Res 229(1):1–20
2. Chekanin VA, Chekanin AV (2012) Effective models of representations of orthogonal resources in solving the packing problem. Inf Control Syst (Informatsionno-Upravlyayushchiye systemy) 5:29–32 (in Russian)
3. Chekanin VA, Chekanin AV (2012) Optimization of the solution of the orthogonal packing problem. Appl Inf (Prikladnaya informatika) 4:55–62 (in Russian)
4. Chekanin VA, Chekanin AV (2012) Researching of genetic methods to optimize the allocation of rectangular resources In: Modern engineering: science and education: proceedings of 2nd international scientific conference. Izd-vo Politekhn. un-ta, SPb. pp 798–804 (in Russian)
5. Chekanin AV, Chekanin VA (2013) Efficient Algorithms for Orthogonal Packing Problems. Comput Math Math Phys 53(10):1457–1465
6. Chekanin AV, Chekanin VA (2013) Improved packing representation model for the orthogonal packing problem. Appl Mech Mater 390:591–595
7. Chekanin VA (2011) The effective solving of the two-dimensional packing task of rectangular objects. J Comput Inf Technol (Vestnik komp'yuternykh i informatsionnykh tekhnologiy) (6):35–39 (in Russian)
8. Chekanin VA, Kovshov EE, Hue NN (2009) Increase of efficiency of evolutionary algorithms at the decision optimization tasks of objects packing. Control Syst Inf Technol 3(37):63–67 (in Russian)
9. Chekanin VA, Kovshov YeYe (2010) Evolutionary-algorithm-based modelling and optimization of the processing steps in industrial production. Tekhnol Mashinostroeniya 3:53–57 (in Russian)
10. Dyckhoff H (1990) A typology of cutting and packing problems. EJOR 44:145–159
11. Garey M, Johnson D (1979) Computers intractability: a guide to the theory of NP-completeness. W.H.Freeman, San Francisco, 338 p
12. Lodi A, Martello S, Monaci M (2002) Two-dimensional packing problems: a survey. Eur J Oper Res 141(2):241–252
13. Martello S, Pisinger D, Vigo D (2000) The three-dimensional bin packing problem. Oper Res 48(2):256–267

14. Mukhacheva EA, Bukharbaeva LY, Filippov DV, Karipov UA (2008) Optimization problems of transport logistics: operating bin stowage for cargo transportation. Inf Technol (Informacionnye Tehnologii) (7):17–22 (in Russian)
15. Philippova AS (2006) Modeling of evolution algorithms for rectangular packing problems based on block structure technology. Inf Technol (Informacionnye Tehnologii). No 6. Appendix. 32 p (in Russian)
16. Washer G, Haubner H, Shumann H (2007) An improved typology of cutting and packing problems. EJOR 183(3):1109–1130

Chapter 6
Designing a Power Converter with an Adaptive Control System for Ultrasonic Processing Units

Vladimir I. Diagilev, Valery A. Kokovin and Saygid U. Uvaysov

Abstract We describe the features of circuit design construction and design of power converters included in ultrasonic process plants. We discuss the results of simulation of an equivalent circuit of the power converter. We then develop information flow diagrams of adaptive control system of the power converter.

Keywords Ultrasound technology · The power converter · Data flows · Data flow diagrams

Introduction

In machinery we find widely used sonication and purification of different parts [1]. Methods of acoustic control in nondestructive testing of engineering products are also widespread. In this case, efficiency increase of ultrasonic processing units (UPU) is an insistent task. There are different ways to increase the efficiency of UPU: using modern methods of constructing the power converters (automatic tuning of generator frequency, control of permissible oscillation amplitude of the piezoelectric transducer (PETs), etc.), improvement in performance of the piezo-electric transducer itself (a rise in sensitivity, efficiency factor and reduction of

V.I. Diagilev (✉) · V.A. Kokovin
The Public Institution of Higher Education of Moscow Region
the International University "Dubna" of Nature, Society and Man,
Branch "Protvino", Severny pas. 9, Protvino 142281, Russia
e-mail: dvi-39@mail.ru

V.A. Kokovin
e-mail: kokovin@uni-protvino.ru

S.U. Uvaysov
NRU HSE, 101000 Moscow, Russia
e-mail: s.uvaysov@hse.ru

© Springer International Publishing Switzerland 2015
A. Evgrafov (ed.), *Advances in Mechanical Engineering*, Lecture Notes
in Mechanical Engineering, DOI 10.1007/978-3-319-15684-2_6

mechanical losses etc.). This article presents a method of constructing the power converter (PC), that includes the adaptive parameter control system of output signals in order to maximize the efficiency of the UPU.

In technological processes, associated with the handling of solids by ultrasound, and at cleaning with high amplitude fluctuations, the main technological parameter is the oscillation amplitude. In [2] it is shown that for ultrasonic transducer (UT) the curve of oscillation amplitude dependence on the input power has a linear (with capacity not exceeding 0.9 kW) and a non-linear plot. With an amplitude increase of input voltage the amplitude of the oscillations or acoustic power increases to a certain limit, the value of which is determined by the fatigue strength of structural elements UT. In the nonlinear regime with increasing of voltage amplitude at the input the mechanical or electrical losses in the UT will also increase. This leads to a decrease in sensitivity, performance and the efficiency factor. In this case, the *allowable vibration amplitude* UT accepts the value at which decrease in sensitivity or efficiency factor does not exceed a predetermined value [2].

It is known [3] that the piezoelectric and magnetostrictive transducers are resonant systems. Maximum power from the PC is given to the load when the generator frequencies of PC are equal to *resonance frequency* of the piezoelectric transducer, when reactive load components (PETs) are compensated and there is only resistance. This factor determines the need of tuning the PC generator frequency to the changing load frequency during the use of UPU.

The Structure of the UPU Power Converter

Based on the requirements for setting PC, it is necessary to ensure the following requirements for an ultrasonic generator (UG):

1. Amplitude of the PETs input voltage should be stable and adjustable.
2. Frequency of this voltage (F) should to be equal to the resonance frequency of the tool attached to the PETs, i.e. it is necessary to adjust F during the operation.
3. A PC Control system should provide a rapid response to PET's output parameters changes.

A PC functional scheme (presented in Fig. 6.1) is proposed for ensuring these requirements. This scheme forms the harmonic output signal during the operating of transistor generator in mode key [4], it makes possible to achieve high efficiency. Moreover, the adaptive PC control system maintains stable amplitude of output voltage (control block (CB-1)) and required resonant frequency (control block (CB-2)). Changing the nominal values of the power high-pass filter HPF elements provides an opportunity further to stabilize the generator output voltage (in a given range) when load changes.

Power Source (Fig. 6.2) is used to provide a generator rectified and smoothed voltage network. This is a single-phase diode bridge with a low-pass filter (LPF). Its output is connected with down-type DC converter (DCC) [5]. It consists of a

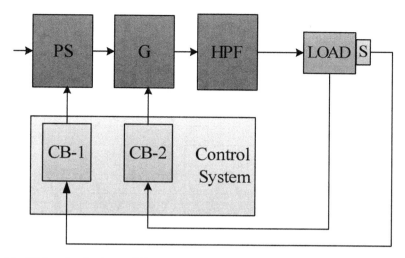

Fig. 6.1 PC functional scheme (*PS* power source, *G* generator, *HPF* high-pass filter, *S* sensor, *CB* control block)

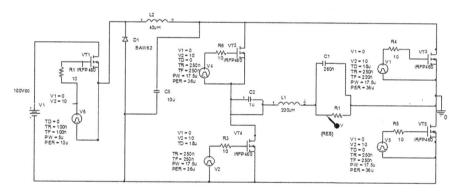

Fig. 6.2 Scheme for PC modeling nodes

series-connected transistor, the choke and the diode, providing energy recovery inductor to the load. This scheme provides:

- Regulation of the supply voltage in the range 0–0.9 Um.
- Reduction of low-frequency output ripple through the use of DCC in the control unit CB-1 pulse width modulation (PWM)—controller.
- Protection against overcurrent and overvoltage.
- Maintenance of the input voltage generator accuracy due to the formation of high-frequency switching transistor DCC.
- PC soft start.

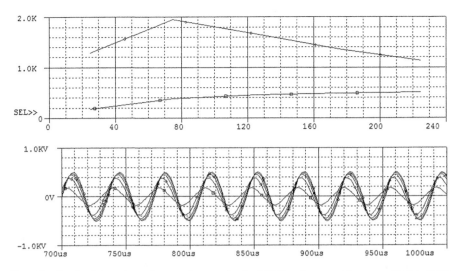

Fig. 6.3 Graph of time dependence voltage at the load $U_l(t)$ and parametric curves $U_l(R_l)$, $P_l(R_l)$

For the PC analysis, an equivalent circuit of power source with DCC and high-frequency generator voltage was made. The power source output terminals are connected with the generator G (Fig. 6.1) with the amplitude of the output voltage equal to the supply and the period, which is determined by the switching speed of the two pairs of cross-bridge transistor. A harmonic voltage filter is used that consists of a resonant tank circuit L1, C1 and current limiting capacitor C2 (Fig. 6.2). Load R1 is connected in parallel with C1 and is active. To determine parameters the simulation of circuits in a resonant mode was carried out.

The simulation results are presented in the form of time dependence voltage at the load when you change the resistance in the range of 25–250 Ω (bottom graph 3). The upper graph shows the parametric curves of load voltage and power (maximum values) dependence on its resistance. Using these curves it is possible to perform the calculation of the parameters of the oscillation circuit and a selection of PCT generator transistors. More details are in [6] (Fig. 6.3).

Selecting Parameters of the PET, Controlled by the Control System

For efficient operation of UPU, as noted earlier in this article, it is necessary to control the *acceptable oscillation amplitude* of the piezoelectric transducer and adjust the frequency of the generator output harmonic signal PC to the PETs *resonance frequency*. In this regard, there is a need to measure the oscillation amplitude of the radiating surface of the oscillating system and measure the current through the PETs, defining its maximum value (resonant mode).

Modern development of microelectronics provides a large selection of tools for measuring the PETs oscillation amplitude. Optical measurement methods [7] and a variety of microelectronic mechanical devices based on MEMS—Technology (Micro-Electro-Mechanical Systems) [8] are widely used. For the following task it is convenient to use MEMS-accelerometer, which permits conversion of oscillation amplitude value to digital output signal. For example, sensor ADXL350 (firm Analog Devices) has a high resolution (13 bits), a wide measurement range and a variety of interfaces. As digital processing of measured parameters is assumed, we choose high-speed interface SPI (Serial Peripheral Interface), which forms synchronous serial data stream. The selected sensor is mounted on PETs body (sensor A, Fig. 6.1).

To maximize the efficiency of using PETs as a part of process unit, current is generally controlled, which measurement can be performed, for example with the use of a transformer current sensor. Another method of controlling a resonant mode is associated with the phase detector that allows selecting the phase relation of PETs voltage and current passing there through [9]. This method is useful for low power radiation or as an additional control to maintain (for example, current control). To control the PETs current resonant value it is necessary to have a transducer amplitude-frequency response characteristic (AFC). Determining methodology is given in [2]. AFC allows specifying the source data for the control system.

Considered list of monitored parameters and requirements for the implementation of monitoring shows a need for an adaptive control system. We define problems for the adaptive system:

- Maintenance of a given output voltage amplitude through generator's power level adjustment (PWM—control). At the same control characteristics can be selected depending on the mode of use of UPU.
- *Allowable* PETs *oscillation amplitude* control through tuning the voltage generator power.
- PETs resonant mode control through adjustment of the oscillator frequency.

Digital processing of the control system is based on the digital matrix—programmable logic integrated circuits (FPGA). Advantages of using this technique is its high matrix performance, convenient implementation of the control algorithm in the form of stream processing, a large number of library functions presented in the form of IP-cores. The big advantage is the ability to use FPGA reconfiguration hardware implementation unlimited number of times. In addition, the development can be done on hardware description languages (Verilog, VHDL), that increases development productivity and allows one to perform functional and timing simulation project.

Modern FPGA incorporates means for effective implementation of the model calculation **data flow**—Digital Signal Processors (DSP), which enables one to develop and implement high-performance control loops PC [10].

Problems of PC lower level digital control, for example, the control process of the power switches at resonant frequency tuning, information obtaining by SPI—interface (sensor vibration amplitude) in real time requires the performance of the FPGA and an order of less than several microseconds.

To ensure the formation of a predetermined load setting mode UPU (including uploading the local network), the microprocessor can be used to network ports.

Determination of the Control System Information Flows

To develop a PC control system it is necessary to analyze the input information flows which are generated at the system's output. Such an analysis is conveniently performed using data flow diagrams (Data Flow Diagrams—DFD), which can be used to develop a specification of requirements in system design. Classical diagram DFD is used for structural analysis and design of information systems. There are several widely used notations DFD [11] with different syntax. Diagrams describe the sources and data flow, represent processes and memory for storing results of processing data streams in the process. These diagrams cannot be directly used in the analysis of information flows control systems (CS) power converters for the following reasons:

- Processes of stream processing must be strictly determined.
- Each process usually runs on a certain condition, therefore, necessary to process input signals to determine the shaping condition logic.

Consider the work of CS at the example of simplified diagram (Fig. 6.4), making it possible to analyze the data flow from external sources, flows processing, logical calculations and destination of calculation results. The diagram is based on data processing units, so called capsules. The capsule consists of two components: the logic condition and processing of data flow. Logic condition is a logical function of the equation, which solution starts the process. The process converts the input data flow to output, generating control signals for other capsules. Wherein, the process algorithm is involved into the capsule (encapsulated). Each capsule can be a hierarchy of nested processes capsules [12].

Sources of input data flow (external entities) CS considered are listed below:

- ADC—analog-to-digital converter in the path of the current control mode resonant probe.
- RR—converter's oscillation amplitude reference register.
- S—control PET sensor of allowable oscillation amplitude.
  Addressee stream processing results:
- G—PC generator.
- DCC—DC converter.
- clk—external event to synchronize all processes.
- str—external event to initialize the PC works.

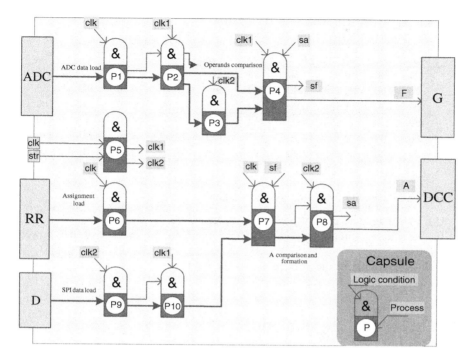

Fig. 6.4 Graph of time dependence

Analyzing the diagram we can see, that control system receives three independent data flow and two events. As a result, two control signals are generated at the output:

- F—generator frequency setting signal for the resonance of PETs.
- A—signal providing a predetermined value generator power through the RR (and thus the acoustic power), the amplitude of the oscillation is controlled riding probe.

Since the digital part is implemented on FPGA CS, the resulting chart with parallel processing of data streams easily implemented in the project, which requires algorithms processes (P1–P10) each capsule described in terms of HDL (e.g., Verilog).

To check the results of the mathematical modeling experiment there was provided an experiment on a physical model of PC. Layout of its capacity of about 500 W power transistors assembled JRF840, with a control system implemented on FPGA family Cyclone III (firm Altera).

A generator oscillatory circuit has the following parameters: L1 = 200 mH, C1 = 230 nF, C2 = 0. Frequency F = 1/2t where t—half-harmonic of the output voltage (t = 20 mS, F = 25 kHz). Tests were conducted at a supply voltage generator Ud = 20 and 30 V. Source impedance Ro = 1,56 Ohm. Table 6.1 shows the results of experiments.

Table 6.1 The results of the simulation and test layout

№	$R_{н}$	U_d	Model				Experiment			
			I_d	I_L	$U_{нм}$	$P_{н.ср}$	I_d	I_L	$U_{нм}$	$P_{н.ср}$
1	25	30	0.60	1.30	28	33	0.80	1.50	50	45
2	50	30	0.82	1.90	52	60	0.95	1.90	75	56
3	100	30	1.62	3.00	93	100	2.00	3.20	92	100
4	150	30	1.95	4.00	124	125	2.40	4.20	85	110
5	200	30	2.30	4.70	147	120	2.50	4.90	155	110
6	100	20	1.50	2.90	66	–	0.70	2.10	33	90

This table uses the following symbols signals:

- Ud, Id-voltage and current power supply generator.
- IL—oscillating circuit current, which is equal to the current of the power transistor.
- Unm—amplitude of the voltage across the load.
- P.av—load power (average value).

Conclusion

Analysis of the data shows a good convergence of the simulated results and the experimental results (the difference does not exceed 25 %). Large deviations of certain values can be attributed to random error.

We have described a scheme for a power converter with improved UPU Stability generator output voltage and the precision of its adjustment, and the control process admissible oscillation amplitude of the tool. The structure of the adaptive system allows an increase in the efficiency of a process plant.

This work was supported by the Russian Foundation for Basic Research (project NK14-07-00422/14).

References

1. Petushko IV (2005) Equipment for ultrasonic treatment—scientific publication. Izd-vo (Andreevskiy izdatel'skiy dom), SPb. 166 p (in Russian)
2. Kazantsev VF (1980) Calculation of ultrasonic transducers for process plants. Izd-vo "Mashinostroenie", M. 44 p (in Russian)
3. Physical basis of ultrasound technology. Pod red. LD Rozenberga. Izd-vo "Nauka", M. 1970. 688 p (in Russian)
4. Kozyrev VB, Lavrushenkov VG, Leonov VP et al (1985) Transistor generators of harmonic oscillations in a key mode. Izd-vo "Radio i svyaz",—M. 192 p (in Russian)

5. Rozanov YuK (1992) Fundamentals of Power Electronics. Izd-vo "Energoatomizdat", –M. 296 p (in Russian)
6. Dyagilev VI, Kokovin VA, Uvaysov SU (2014) Research processes in the circuit transistor ultrasonic generator for process plants. Sovremennoe mashinostroenie. Nauka i obrazovanie: materialyi 4-y mezhdunarodnoy nauchno-prakticheskoy konferentsii. -SPb.: Izd-vo Politehn. un-ta, pp 1250–1257 (in Russian)
7. The Makarov LO (1983) Acoustic measurements during ultrasonic technology. Izd-vo "Mashinostroenie"—M. 56 p (in Russian)
8. Vijay K, Varadan KJ, Vinoy KA, Jose RF (2004) MEMS and their applications. Izd-vo "Tekhnosfera", M. 526 p (in Russian)
9. Khmelev VV, Pedder SN, Tsyganok RV, Barsukov RV (2009) The features investigation of piezoelectric transducers. In: International conference and seminar on micro/nanotechnologies and electron devices. EDM'2009: conference proceedings—Novosibirsk: NSTU, pp 233–241
10. Guo J, Edwards SH, Borojevic D (2002) Implementing dataflow-based control software for power electronics systems—In: Proceedings of IEEE 9th workshop on computers in power electronics (COMPEL)
11. Marka DA, MakGouen KL (1993) Structured analysis and design technique (SADT). Izd-vo "MetaTekhnologiya", M. 240 p (in Russian)
12. Kokovin VA, Evsikov AA (2014) Streaming techniques use management automation in manufacturing plants. Sovremennoe mashinostroenie. Nauka i obrazovanie: materialyi 4-y mezhdunarodnoy nauchno-prakticheskoy konferentsii. -SPb: Izd-vo Politehn. un-ta, pp 1275–1284 (in Russian)

Chapter 7
Features of Multicomponent Saturation Alloyed by Steels

Sergey G. Ivanov, Irina A. Garmaeva, Michael A. Guriev, Alexey M. Guriev and Michael D. Starostenkov

Abstract In this work we consider the mechanism of formation and structure of multicomponent diffusive boride coverings on steels.

Keywords Chemical heat treatment · Hardening · Borating

Introduction

In modern production of tools, materials are required to have high surface strength, abrasion resistance and other characteristics. Diffusion coatings based on boron have a set of desired properties: high surface hardness, wear resistance, corrosion resistance and hardness at elevated temperatures [1–5]. In this paper we study the influence of the medium and process parameters of process hardening on the microstructure and properties of boride layers on steel alloys for the production of stamps. As objects of study we have selected the following steels: 5HNV, 5HNVM and X12M.

S.G. Ivanov · I.A. Garmaeva · M.A. Guriev · A.M. Guriev · M.D. Starostenkov (✉)
Altai State Technical University after I.I. Polzunov, Barnaul, Russia
e-mail: genphys@mail.ru

S.G. Ivanov
e-mail: serg225582@mail.ru

I.A. Garmaeva
e-mail: gurievam@mail.ru

M.A. Guriev
e-mail: gurievma@mail.ru

A.M. Guriev
e-mail: gurievam@mail.ru

© Springer International Publishing Switzerland 2015
A. Evgrafov (ed.), *Advances in Mechanical Engineering*, Lecture Notes
in Mechanical Engineering, DOI 10.1007/978-3-319-15684-2_7

Experiment

The surface structure of the hardened steels we studied are mainly formed of three chemical elements, such as iron, boron and carbon. Iron is an essential element of the saturable alloy, boron—the main alloying element on the surface, the carbon is present in an amount introduced into the steel. Images of microstructure for borated and complexity of hardened boron and titanium of steels for the production of stamps are presented in Figs. 7.1 and 7.2.

Equilibrium in the system Fe–B–C in the process of hardening leads to the formation of three stable borides: FeB, Fe_2B and Fe_3B [6, 7]. Also noted is the presence of a metastable boride $Fe_{23}B6$, which has a cubic structure. It should be noted that unlike carbon the solubility of boron in iron (γ), (α) and (σ) is very limited [6, 8].

In addition to the binary phases FeB, Fe_2B, Fe_3B, Fe_3C, in the investigated steels we observed availability ternary phases: $Fe_3(C, B)$ and $Fe_{23}(C, B)_6$ and ferrite.

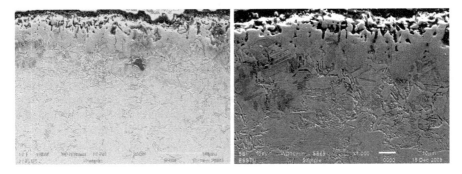

Fig. 7.1 The microstructure of boride layer on the steel for the stamp

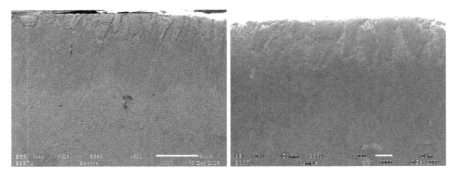

Fig. 7.2 The microstructure of the diffusion layer obtained by simultaneous saturation of boron and titanium on the steel for the production of stamps

Carbides Fe_3C and $Fe_{23}C_6$ are isomorphic to the borides Fe_3B and $Fe_{23}B_6$ therefore easily formed corresponding to the phase of ternary symmetry, namely, $Fe_3(C, B)$ and $Fe_{23} (C, B)_6$.

We expect the possible presence of phases of Boron and Carbon, namely, B_4C, $B_{13}C_2$, B_mC_n, as well as other options for the presence of boron carbide. However, in this study, boron carbides were not found. This is due to the presence of narrow areas stability of boron carbide and a small amount of carbon in the surface layers of steel as a result of decarburization at saturation of boron [9].

Phase composition was diagnosed on diffraction patterns obtained in two ways: (1) X-ray diffraction analysis, and (2) electron diffraction microscope.

Borated cementite $Fe_3(C, B)$, and normal cementite Fe_3C differ greatly morphologically. Prior to the penetration of boron in the crystal lattice can be seen clearly lamellar cementite structure (Fig. 7.3a). After penetration of boron and education borated cementite, the morphology changes—instead of the perfect plates we obtained fragmentary structure (Fig. 7.3b).

Borated plates of cementite are crushed and fragments of these plates are clearly visible in Fig. 7.3. During saturation of steels by boron the volume of fraction cementite increases. This can be explained by observing that diffusion boron in cementite binds an additional share of iron. Availability borated cementite in all studied steels is also confirmed by X-ray analysis.

In the case of a complex saturation with boron and titanium the diffusion layer undergoes slight changes (Fig. 7.2). It is formed in a two-layer structure, the upper part of which consists of columnar crystals of borides of iron, doped titanium, and the lower (dark) portion is a mixed compounds of iron and titanium form Fe_xTi_y (B_n, C_m). Furthermore, in this zone is present up to 22 % of titanium, and about 0.47 % carbon. Such a distribution of elements is likely to occur as a result of mutual diffusion of carbon from the depths of metal and the diffusion of carbon from the surface of metal as a result of its diffusion under the action of boron.

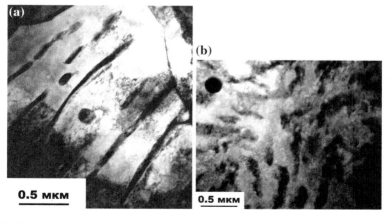

Fig. 7.3 The microstructure of a conventional cementite (**a**) and borated cementite (**b**)

Mechanical properties of this zone is demonstrated to increase the microhardness of the needles of boride of the order of 17–22 % and a transition zone of about 25–35 %, which in turn affect the increase in durability of hardened boron and titanium of steel. The remainder of the coating and transition zones do not differ from the boride coating.

In the case of boriding steel X12M, the diffusion layers have a high content of alloying elements, the rate of diffusion of boron and chromium significantly slowed. The results of elemental analysis of the surface of the hardened samples showed that the chromium content remained practically unchanged with respect to the steel, not subjected to treatment. The maximum boron concentration in the diffusion layer does not exceed 18 %. In the case of a simultaneous diffusion of boron, titanium and chromium the elemental composition of the coating varies over a wide range: the chromium content is reduced from 12.04 to 10.5 %, while the titanium concentration increases from 0.16 to 14.72 %.

Conclusion

The highest concentration of titanium is achieved primarily in the surface layer on the defects of crystal structure, on the grain boundaries and between borides, a small amount of titanium is alloying of iron borides. For a more complete picture of the distribution of elements and the phase structure of the diffusion layer, one requires research methods of X-ray diffraction and electron diffraction, which allows one to establish with a high degree of reliability of phase composition, crystal structure and chemical formulas of individual phases of the boride coating.

Acknowledgments This work was supported in part by the Russian Foundation for Basic Research (Project No. 13-08-98107 "r_Sibir'_a") and a grant from the President of the Russian Federation Treaty 14.Z56.14.656-MK.

References

1. Guriev AM, Ivanov SG, Greshilov AD, Zemlyakov SA (2011) The mechanism of formation of boride needles with someone diffusion of saturating the complex chromoborating plasters. Met. Finish.: technol. equip. tool. (Obrabotka metallov: technologya, oborudovanie, instrumenti) 3:34–40 (in Russian)
2. Guriev AM, Lygdenov BD, Maharov DM, Mosorov VI, Chernykh EV, Gurieva OA, Ivanov SG (2005) Features of structure formation of the diffusion layer on the cast steel with chemical-thermal treatment. Fundam. probl. of mod. mater. (Fundamentalnie problemy sovremennogo materialovedenia) 2(1):39–41 (in Russian)
3. Ivanov SG, Garmaeva IA, Guriev AM (2012) Features diffusion of boron atoms and chromium at two-component surface saturation St3 steel. Fundam. probl. of mod. mater. (Fundamentalnie problemy sovremennogo materialovedenia) 9(1):86–88 (in Russian)

4. Guriev AM, Ivanov SG (2011) The mechanism of diffusion of boron, chromium, and titanium during multicomponent saturating the surface of the iron alloys. Fundamental problems of modern materials (Fundamentalnie problemy sovremennogo materialovedenia) 8(3):92–96 (in Russian)
5. Guriev AM, Ivanov SG, Lygdenov BD, Vlasova OA, Kosheleva EA, Guriev MA, Garmaeva IA (2007) Influence of parameters chromborating on the structure of steel and mechanical properties of the diffusion layer. Polzunovsky Vestnik 3:28–34 (in Russian)
6. Lygdenov BD, Garmaeva IA, Popova NA, Kozlov EV, Guriev AM, Ivanov SG (2012) The phase composition and the defective condition of gradient structures at borated steel 20L, 45, 55 and 5HNV. Fundam. probl. of mod. mater. (Fundamentalnie problemy sovremennogo materialovedenia) 9(4–2):681–689 (in Russian)
7. Guriev AM, Kozlov EV, Zhdanov AN, Ignatenko LN, Popova NA (2001) Change of phase composition and mechanism of formation of the structure of the transition zone in the temperature cycling boriding ferrite-perlite steel. In: Proceedings of the higher educational institutions, Physics, vol 2, p 58 (in Russian)
8. Guriev AM, Kozlov EV, Crimskikh AI, Ignatenko LN, Popova NA (2000) Change of phase composition and mechanism of formation of the structure of the transition zone in the process saturating of the carbon and boron of temperature cycling the ferrite-perlite steel. In: Proceedings of the higher educational institutions, Physics, vol 43, p 60 (in Russian)
9. Ivanov SG, Guriev AM, Starostenkov MD, Ivanova TG, Levchenko AA (2014) Special features of preparation of saturating mixtures for diffusion chromoborating. Russian Phys J 57 (2):266269

Chapter 8
New Modelling and Calculation Methods for Vibrating Screens and Separators

Kirill S. Ivanov and Leonid A. Vaisberg

Abstract This work reveals the authors' experience in developing a fast-acting computational modelling algorithm and calculation of devices for vibrational size classification of ores, solid waste and other bulk materials. It is shown that adaptation of several analytical models to computing allows creating a new, less resource-intensive approach, as compared to widespread multi-functional computational methods (including DEM). This algorithm, based on problem-oriented models, requires neither preliminary calibration nor adjustment, which additionally reduces its deployment time and application costs. Moreover, the algorithm enables parcel-wise improvements, as illustrated by introduction of an advanced openings-passing model and a layer-based vibrational separation sub-model.

Keywords Screening · Separation · Numerical modelling

Introduction

Vibrational screening is a widely used size classification method for bulk materials. The nature of this process in open and, especially, closed fragmentation cycles significantly affects energy efficiency of disintegration. When processing non-metallic minerals and coal, vibrational screening ensures high commercial quality of final products. Size classification of municipal solid waste also represents a critical operation in material preparation for further processing. During the long application history of vibrating screens, three fundamentally different types of approaches to calculating respective screening devices have been developed. Empirical approaches are based on direct generalization of the available experience

K.S. Ivanov (✉) · L.A. Vaisberg
REC "Mekhanobr-Tekhnika", St. Petersburg, Russia
e-mail: ivanoff.k.s@gmail.com

L.A. Vaisberg
e-mail: gornyi@mtspb.com

© Springer International Publishing Switzerland 2015
A. Evgrafov (ed.), *Advances in Mechanical Engineering*, Lecture Notes in Mechanical Engineering, DOI 10.1007/978-3-319-15684-2_8

and simple formulas with experimental coefficients; phenomenological approaches are based on hypotheses regarding various structures of a particular process, often referred to as screening kinetics; and numerical approaches were formed at the end of the last century, when computing capacities became sufficient to carry out detailed modelling of physical processes with given mechanical properties.

The discrete element method [1] and other common universal modelling approaches for the behavior of granular materials currently used to study vibrational screening require extensive computing resources. Detailed numerical experiments may take months, which eliminates the benefits a computer simulation has as compared to a full-scale experiment. Therefore, the development and improvement of well-known analytical models may significantly simplify the process of developing new equipment. Moreover, when combined with such advanced methods of numerical optimization as, for example, the particle swarm optimization [2], these improved approaches will enable automatic determination of optimal design and process-related parameters for sieve classification devices.

Model of Dry Vibrational Screening Processes

The systematic approach proposed by Mekhanobr Institute and most fully represented in [3] has been developed for modelling the screening of a thick layer of coarse bulk material transported over the screening surface at a constant average speed. The authors of [3] derived a most general differential equation determining the screening kinetics as follows:

$$\frac{d\varepsilon_D}{dy} = \frac{u\phi}{v}\left(1 - \frac{D}{d_0}\right)^2 P_{D,y}(0)(1 - \varepsilon_D),$$

where ε_D—undersize recovery for narrow class D, v—material layer velocity over the screen, u—particle penetration rate through the openings in the screening surface, d_0—diameter of screen openings, and $P_{D,y}(0)$—the proportion of diameter D particles in the contact area under consideration.

Assuming the material in the layer is uniformly mixed, this equation transforms into the central differential equation from classical model [4]. In the case of absolute segregation and linear cumulative particle size characteristics of the material, the equation may be integrated analytically and reduced to the equation of

$$\gamma = \frac{\beta y}{y + \frac{Q_0\beta}{u\phi ac}},$$

which coincides with those obtained experimentally in [5]. Here, γ represents the undersize yield of the material, β is the parameter characterizing the gradient of the cumulative particle size characteristic, Q_0—feed rate, and c—bulk density of the material. It should be emphasized that, as compared to the currently most common numerical approaches to modelling the behavior of granular medium, this method requires extremely low computational resources.

The Improved Approach

The improved approach to modelling the screening process for coarse bulk materials presented below is based on the extensive experience of REC "Mekhanobr-Tekhnika" in the design and manufacture of vibrating screens. The program developed using this approach has the following input parameters: grain-size distribution of the material, represented as a mixture with given proportions of narrow grain-size classes of user-defined diameters d_i; coefficients of friction between material particles and with the screening surface; screen feed rate (Q_0); screen parameters, such as its length (l), width (a), opening diameter (d_0) and the open area (ϕ); screen oscillation characteristics (horizontal and vertical components may be set separately in an arbitrary form) or, similarly to the basic model, the vibrational displacement rate of the material layer on the screen surface (v); as well as the velocity of material particle penetration through screen openings (selectable from a special table of experimental data). This program outputs recovery values for all predetermined narrow classes, distribution of these grain-size classes in the material layer over the entire length of the screening surface, as well as all derivative characteristics such as recovery efficiency for a selected class or a set of classes, product yields, etc.

Since the differential equation in the basic model is generally not analytically integrable, the improved approach was originally formulated in a form suitable for computer-based calculations. The material layer on the screening surface was represented as a set of spatial cells (see Fig. 8.1).

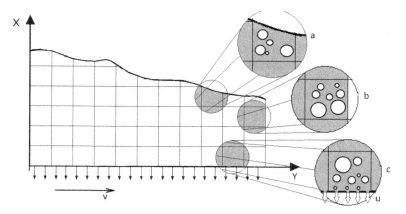

Fig. 8.1 Longitudinal section of material layer on the screen, cell types: **a** surface layer cell, **b** the material thickness cell, **c** cell in the contact area between the material and the screen

For each cell (number j vertically from the bottom of the column, with number k as counting from the feeder), the following parameters are set: $F_{k,j}(d_i)$—mass fraction of narrow class d_i in the cell, $M_{k,j}$—gross weight of all the material in the cell.

This material distribution into cells, which facilitates computations, also enabled developing a modular architecture for the program. The approach is based on three relatively independent sub-models regarding displacement, screening and redistribution of particles within the material layer. Focusing on specific modelling goals and objectives, each user may change individual models and related sub-programs in order to achieve a reasonable balance between resource intensity and calculation accuracy or even use a new sub-program based on a different model without changing the remaining code.

Vibrational displacement was included in the basic model only as a parameter v—velocity of the material layer on the screening surface. With that, the connection between vibration parameters and material shift along the screening surface was taken into account only indirectly, which in some cases may complicate engineering calculations. In order to solve the general problem, a sub-program was developed simulating vibrational displacement of a solid body on a rough inclined vibrating plane. The program was based on the methods presented in [6]. The average displacement velocity in this case was assumed to be equal to the material layer displacement velocity under the same conditions.

In accordance with the basic model, mass outflow through a screen segment number k for each class may be calculated by the formula:

$$q_k(d_i) = P(d_i)uF_{k,0}(d_i)\Delta ya,$$

where $P(d_i)$ is a coefficient determining the proportion of particles with diameter d_i on the screening surface that will pass through the screen, conventionally referred to as the penetration probability of narrow class particles; while u, $F_{k,0}(d_i)$, Δy and a correspond to the rate of material penetration through screen openings, mass fractions of different classes in granular material cells, horizontal cell size and screening surface width respectively.

Similarly to most of the preceding analytical approaches, the basic model is based on the Gaudin penetration probability. However, the classical Gaudin formula does not assume the possibility of simultaneous passage of several particles of the same or different sizes through a single screening surface opening, which may result in a somewhat understated screening intensity for finer grain-size classes. The model presented in this paper has been refined using an advanced formula by Pelevin [7]. If the Gaudin penetration probability for a particle with diameter is

$$P^{d_0}(d_i) = \varphi\left(1 - \frac{d_i}{d_0}\right)^2,$$

the refined Pelevin probability may be presented in the compact form of

Table 8.1 Results of full-scale and numerical modelling using the presented approaches

No	Source material classes (mm)	Contents of grain-size classes (%)	Screen weight output (t/h)	Sieve cell size (mm)	Recovery			
					Experimental data	Basic model	Improved screening	Improved separation
1	−0.071 + 0	1.67	1.76	0.630	0.83	0.97	0.96	0.98
	−0.18 + 0.071	11.41			0.91	0.97	0.96	0.95
	−0.315 + 0.18	8.58			0.88	0.97	0.95	0.86
	−0.63 + 0.315	16.74			0.53	0.84	0.82	0.61
	+0.63	61.61			0.00	0.00	0.00	0.00
2	−0.071 + 0	1.26	0.63	0.630	1.00	0.98	0.97	0.99
	−0.18 + 0.071	11.12			1.00	0.98	0.97	0.97
	−0.315 + 0.18	8.84			1.00	0.98	0.97	0.90
	−0.63 + 0.315	17.12			0.69	0.95	0.93	0.72
	+0.63	61.66			0.00	0.00	0.00	0.00
3	−0.071 + 0	1.24	0.54	0.315	1.00	0.97	0.95	0.96
	−0.18 + 0.071	11.17			0.99	0.96	0.95	0.90
	−0.315 + 0.18	10.30			0.74	0.93	0.90	0.75
	−0.63 + 0.315	16.56			0.00	0.00	0.00	0.00
	+0.63	60.73			0.00	0.00	0.00	0.00
4	−0.071 + 0	1.53	1.73	0.315	0.89	0.96	0.94	0.95
	−0.18 + 0.071	11.98			0.87	0.95	0.93	0.88
	−0.315 + 0.18	8.59			0.61	0.77	0.74	0.61
	−0.63 + 0.315	17.02			0.00	0.00	0.00	0.00
	+0.63	60.87			0.00	0.00	0.00	0.00

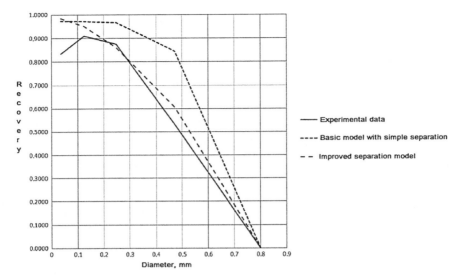

Fig. 8.2 Recovery curves by grain-size classes for experiment 1

$$P(d_i) = P^{d_0}(d_i) + \sum_{j,d+d_i \le d_0} F_{k,0}(d_j) P^{d_0}(d_j) P^{d_0-d_j}(d_i).$$

Here, the items in the sum correspond to the conditional penetration for a particle with diameter d_i, assuming the penetration of particles with diameter d_j in the remaining gap of $d_0 - d_j$ for j values, for which $d_j + d_i \le d_0$.

Processes occurring inside the material layer during its vibrational transportation may significantly influence sieve classification, since they determine which particles come in contact with the screening surface. Various ways to describe these processes and account for them in vibrational screening modelling have been considered in many studies [4, 8]. Specific features of the model presented in this paper enable flexible implementation of virtually any ideas on modelling the behavior of particles inside the material layer. The basic version of the model treated separation as a uniform process through its forced averaging with a preset proportionality of the distribution function for grain-size vertical distribution along the section with the function corresponding to a completely sorted distribution. The above vibration segregation treatment method has, however, a severe drawback—essentially heterogeneous materials in this modelling are sorted at the same rate as practically homogeneous materials. In this regard, a refined model was added to the algorithm, representing separation as part of the material exchange between computational cells with fine particles sieved through a layer of large particles, as if through a screen (Table 8.1 and Fig. 8.2).

Conclusion

The approach to vibrational screening modelling presented in this paper has several advantages as compared to other currently existing numerical and common analytical methods. It enables high-accuracy modelling without time-consuming setup, calibration and preliminary experimental preparations and may be applied for a vast array of devices. The computer software with a friendly user interface developed on the basis of the approach allows designers to evaluate performance of the designed device by its initial parameters. Moreover, low computing resource intensity of the program enables its application in combination with stochastic optimization methods for preliminary evaluation of optimal parameters for designed screens and screens separators.

References

1. Cundall PA, Strack ODL (1979) A distinct element model for granular assemblies. Geotechnique 29:47–65
2. Kim T-H, Maruta I, Sugie T (2010) A simple and efficient constrained particle swarm optimization and its application to engineering design problems. J Mech Eng Sci 224(C2):389–400
3. Vaisberg LA, Rubisov DG (1994) Vibrational screening of bulk materials. Mekhanobr, Leningrad, p 47
4. Ferrara G, Preti U, Schena GD (1988) Modelling of screening operations. Int J Miner Process 22(1–4):193–222
5. Abramovich IM (1935) Some regularities of screening process. XV years in the service of socialist construction. Jubilee volume of Mekhanobr, Leningrad–Moscow, GRGTL, pp 367–410
6. Blekhman IIII, Dzhanelidze GY (1964) Vibrational displacement. Nauka, Moscow, p 410
7. Pelevin AE (2011) Probability of particles passing through the sieve openings and separation process in vibrational screening devices. Izvestiya vuzov. Min J (1):119–129 (in Russian)
8. Subasinghe GKNS, Schaap W, Kelly EG (1990) Modelling screening as a conjugate rate process. Int J Miner Process 28(3–4):289–300

Chapter 9
Remote Monitoring Systems for Quality Management Metal Pouring

Vladimir F. Minakov and Tatyana E. Minakova

Abstract In this chapter we reveal the most influential quality on processes of production of details of car production at the technological stage to be molding of cast iron in forms, A scheme of remote monitoring of provision of a filling ladle is developed for ensuring quality of filling of metal during production of metal products. The device of remote control of positioning of a ladle with the melted metal is developed. The method of decomposition of linear optical spectral characteristics of objects offers a harmonious row. The characteristics of reflection of objects of the spectrophotometer created by the Swiss firm X-Rite are investigated. Comparison of experimental data and results of submission of characteristics by Fourier's ranks and linearity allows a harmonious row to be executed. An increase of accuracy of decomposition in comparison with Fourier's number is proved. For metrological levels of errors the required number of members of a number of decompositions is defined.

Keywords Mechanical engineering · Automation · Quality control · Monitoring · Optical spectrum

Introduction

Nowadays productions of machine-building enterprises in wide circulation utilizes technologies of remote control and management. It is caused, first, by features of technological processes demanding high temperatures (to 1,500 °C) at production of

V.F. Minakov (✉)
Saint Petersburg State University of Economics (SPbSUE), Saint Petersburg, Russia
e-mail: m-m-m-m-m@mail.ru

T.E. Minakova
National Mineral Resources University (NMRU), Saint Petersburg, Russia
e-mail: t.e.minakova@mail.ru

© Springer International Publishing Switzerland 2015
A. Evgrafov (ed.), *Advances in Mechanical Engineering*, Lecture Notes in Mechanical Engineering, DOI 10.1007/978-3-319-15684-2_9

details of cars and devices a method of molding [11], high speeds at their processing by milling and other methods of machining, etc. [19]. Such parameters of modes of processing exclude possibility of direct contact of the personnel with preparations for measurement of their sizes, control of indicators of quality, etc. [16]. Secondly, remote control and automatic control become unique ways at automation of productions [9, 14]. Automation becomes guarantee of competitiveness of the enterprises. Here productivity, quality of products, power efficiency [12, 13, 17] become important factors of the competition.

A rather widespread problem of monitoring of quality in productions is remote nondestructive control of structure, existence, positioning of details, knots, cars, materials and other objects [8]. This task in control systems of moving and remote objects contact influences what is excluded, especially in actual technological conditions. In mechanical engineering such influences treat temperature (for the melted metals), mechanical (limited by strength limits for hi-tech productions of microelectronics), dimensional (for nanotechnologies, etc.). And such restrictions are critical for conveyor productions.

Remote positioning, identification and control of parameters of objects in mechanics is made on their spectral characteristics of reflection or radiation. The analysis of known ways of remote control of spectral characteristics of objects [3] allows us to establish that existing decisions are based on registration of spectral characteristics, that is measurement of dependences of intensity of the reflected signals (or radiations) from wave length in the range from 200–400 to 800–1,000 nm. Such ways allow one to solve an objective, however, the cost of spectrographs, spectrometers, the spectrophotometers, allowing us to measure intensity of ranges such as wave length, is so high that costs of their acquisition and installation in technological processes are done by production economically inefficient, and production—noncompetitive and illiquid [13, 14].

Therefore, remote control requires a choice of cheaper sensors with the characteristics of sensitivity providing the maximum sensitivity in all range of spectral characteristics. For this purpose comparison of hundreds of couples of data on their sensitivity, and for all let-out types of sensors, is required. The task becomes labor-consuming for the decision. Such technique is unacceptable for practical use.

Research Objective

The purpose of our research is a search for a method of improvement of system of monitoring of quality of production of machine-building production by way of nondestructive control on the basis of identification of optical spectral characteristics of objects.

Solution of a Task

In existing technology of molding of preparations of piston rings, observance of demanded parameters of filling (height of a stream of the melted metal, weight speed of filling) is provided with organoleptic control (visually) the metal caster that means influence of a human factor on a technological process. Depending on the physical and emotional condition of the pourer there is also a stability of parameters of filling. In particular, the demanded height of a stream of filled-in metal at production of piston rings makes 225 mm (with the admission of deviations in 25 mm) from an edge of the top molding to a ladle sock with the melted metal. In case of a deviation of height of positioning of a ladle towards increase, there occurs destruction by a stream of metal of a sandy form, and filled-in metal follows from a forming pile. Besides, there is a suction of air of the surrounding atmosphere in a filling form that finally conducts to receiving defective or low-quality production. As labor and material resources are spent for production of the rejected castings, profitability of production becomes low, and the enterprise—noncompetitive.

In case of fall of height of a stream of less demanded size, there is no sufficient pressure of filled-in metal in forms and as a result, there are defects, gas sinks, nonmetallic inclusions in the form of particles of a sandy form. Such instability of technological process also negatively affects quality of production, and also its prime cost.

Figure 9.1 presents a process of positioning of a ladle when molding metal and its organoleptic control on the basis of perception by organs of vision of the pourer with excess of admissible height, and respectively—a characteristic defect: destruction of a filling form and metal effluence.

Fig. 9.1 Destruction of a filling form at the overestimated provision of a ladle with metal

For the solution of a problem of remote monitoring an analysis of known ways of control in technological processes [4–7, 15, 20] is made. It allowed us to establish that existing technological decisions are based on registration of spectral characteristics, that is measurement of dependences of intensity of the reflected signals (or radiations) from wave length in the range from 200–400 to 800–1,000 nm. Such a way allows us to solve an objective, however, the cost of spectrographs, spectrometers, the spectrophotometers, allowing us to measure intensity of ranges as wave length, is so high that costs of their acquisition and installation in technological processes are done by production economically inefficient, and production—noncompetitive and illiquid.

A more economic way—use of cheap photodiode sensors—is offered. For justification of such a way, measurement of intensity of the reflected optical signal for a sample was executed. The object was executed from cast iron. (In [10]—Fig. 9.2, pilot studies are executed with use of the spectrophotometer of the Colormunki model made by the Swiss firm X-Rite, and DispCal GUI software in the Windows operating system).

Considering the feature of spectral characteristics consisting in growth of intensity (and sensitivity for sensors) at the beginning of the range of lengths of waves, and also decreases it on the right border of range, it would be possible to spread out in a row Fourier [18]. However the fact of discrepancy of intensity of reflection on borders of range of a spectral characteristic pays special attention. And the inequality of initial and final values of functions subject to decomposition is violation of a condition of Dirichlet to which functions for decomposition used in a row have to answer to Fourier. For permission of a contradiction we will prove the theorem.

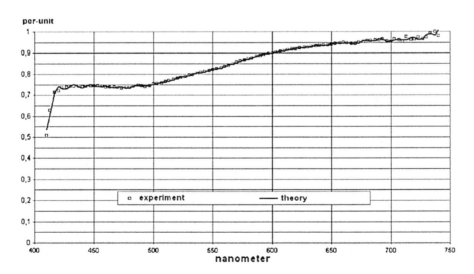

Fig. 9.2 Spectral characteristic and its decomposition in a row

The Theorem of Decomposition of a Linear Acyclic Function—A Harmonious Row

Any acyclic function $f(x)$ on an interval can be spread out in a row [1]:

$$f(x) = a_0 + b_0 \cdot (x - x_1) + \sum_{m=1}^{\infty} (a_m \cdot \cos m \cdot x + b_m \cdot \sin m \cdot x) \qquad (9.1)$$

where: a_0, a_m, b_m—coefficients of a number of Fourier:

$$a_0 = \frac{1}{2 \cdot \pi} \int_{-\pi}^{\pi} f_{\ni}(x)\, dt, \qquad (9.2)$$

$$a_m = \frac{1}{\pi} \int_{-\pi}^{\pi} f_{\ni}(x) \cdot \cos m \cdot x\, dt, \quad m = 1, 2, .., \qquad (9.3)$$

$$b_m = \frac{1}{\pi} \int_{-\pi}^{\pi} f_{\ni}(x) \cdot \sin m \cdot x\, dt, \quad m = 1, 2, .., \qquad (9.4)$$

b_0—coefficient:

$$b_0 = [f(x_2) - f(x_1)]/(x_2 - x_1). \qquad (9.5)$$

For the proof we will transform a decomposition formula in a row as follows:

$$f(x) - b_0 \cdot (x - x_1) = a_0 + \sum_{m=1}^{\infty} (a_m \cdot \cos m \cdot x + b_m \cdot \sin m \cdot x). \qquad (9.6)$$

We will notice that at x_2

$$f(x_2) = b_0 \cdot (x_2 - x_1) + f(x_1), \qquad (9.7)$$

it isn't equal $f(x_1)$. At the same time values

$$f(x_1) - b_0 \cdot (x_1 - x_1) = f(x_1), \qquad (9.8)$$

$$f(x_2) - b_0 \cdot (x_2 - x_1) = b_0 \cdot (x_2 - x_1) + f(x_1) - b_0 \cdot (x_2 - x_1) = f(x_1) \qquad (9.9)$$

are equal each other.

Therefore, for function

$$f(x) - b_0 \cdot (x - x_1), \tag{9.10}$$

first, Dirichlet's condition is met, secondly, the right part—is Fourier's number. Thus, the theorem is proved.

For use of the theorem of decomposition in linearity—a harmonious row in relation to a spectral characteristic with experimentally received values $f_\ni(\lambda)$, we will execute replacement of variables, in particular wave lengths λ on

$$x = (\lambda - \lambda_{\min}) \cdot 2 \cdot \pi / (\lambda_{\max} - \lambda_{\min}). \tag{9.11}$$

Now for decomposition of a linear spectral characteristic—a harmonious row is enough to execute the following algorithm:

- to determine by experimental data

$$b_0 = [f_\ni(\lambda_{\max}) - f_\ni(\lambda_{\min})]/(2 \cdot \pi) \tag{9.12}$$

- to pass from experimental data to function:

$$f_\ni(x) = f_\ni(\lambda) - b_0 \cdot (\lambda - \lambda_{\min}) \cdot 2 \cdot \pi / (\lambda_{\max} - \lambda_{\min}). \tag{9.13}$$

- to determine row coefficients by formulas (9.2, 9.3, 9.4).

Thus, coefficients linearly—a harmonious row (9.1) are identified.

We will apply the offered decomposition to research of dependence of an error of approximation of a range linearly—a harmonious row in function from number of the considered members of a row. The constant component of a Fourier number makes $a_0 = 0.61468$ relative units (per-unit) when rationing intensity of the reflected range to single value at its maximum. The tangent of angle of an inclination linear component is equal to $b_0 = 0.0014346$. Results of decomposition are in a row presented in Table 9.1.

The results given in Table 9.1, lead to the following important conclusions. The format of dot representation of ranges of objects for a hundred values of lengths of waves can be replaced with couples of tabular data, then linearly—harmonious decomposition in a row with two parameters constants, and also two amplitude values of harmonious components. Thus the error won't exceed 4 %. Linearly—the harmonious way of decomposition can be applied in metrological tasks as its error is sufficient for providing a class of measuring devices. The increase in number of considered harmonicas, for example, to six, provides a class of accuracy of representation of the ranges, not exceeding 1 (that is having the given error to 1 %), to eleven harmonicas—for receiving an error to 0.5 %. Therefore, the developed technique of identification of spectral characteristics of objects and sensors for their registration in systems of monitoring of quality of production provides the demanded accuracy.

Table 9.1 Errors of decomposition of a spectral characteristic

Number of harmonicas of a Fourier number	Amplitudes of harmonicas		Average error (%)
	Cosinusoidal	Sinusoidal	
1	−0.036694854	−0.007544815	3.96
2	−0.006895065	−0.04140526	2.52
3	0.025988274	−0.000127576	1.87
4	−0.001988306	0.019622916	1.34
5	−0.012935454	−0.001251574	1.15
6	0.000315431	−0.010941673	0.91
7	0.0082879	0.001011139	0.8
8	−0.000775432	0.006722954	0.76
9	−0.006636308	−0.001216115	0.7
10	0.000889407	−0.006536407	0.6
11	0.005433259	0.000836082	0.5
12	−0.000551821	0.003558956	0.47
13	−0.003157161	−0.000936825	0.45
14	0.001267087	−0.003209484	0.44
15	0.003650987	0.00101817	0.40
16	−0.001075222	0.002485946	0.38
17	−0.002099956	−0.00100592	0.42
18	0.000865523	−0.00302113	0.43
19	0.003362577	0.000630892	0.39
20	−0.00118664	0.002683981	0.37
21	−0.002037598	−0.001332573	0.37
22	0.001264398	−0.002396875	0.38
23	0.002941593	0.001179685	0.45
24	−0.001501723	0.002730271	0.38
25	−0.002271678	−0.001139593	0.361

In Fig. 9.3 the scheme of the automated control of height of an arrangement of a filling ladle [2 14] is submitted.

Control is based on registration of radiation of a surface of the melted metal on a spectral characteristic and management of received information of a light board—a line of light-emitting diodes. The sensors which are most sensitive to a range of radiation are of melted metal and take place at height of demanded provision of a ladle when filling fusion from it in piles of filling forms—at the height of 200 mm from an edge of a funnel of the top pile that provides control of height of a stream in the range of 200–250 mm. In the horizontal plane optimum distance of fastening the device of remote control from a ladle makes 300–400 mm. At smaller distance there is a danger of hit of splashes of the melted metal with a temperature 1,430 °C on sensors. At bigger distance, sensitivity of the monitoring system decreases. Light-emitting diodes

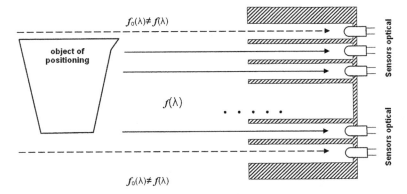

Fig. 9.3 Scheme of remote control of provision of a filling ladle

of indication are located on the same panel. Sampling of an arrangement of sensors and indicators on height makes 20 mm that is quite enough for ladle positioning at the height of 200–250 mm from a pile of filling forms.

Conclusion

Remote monitoring in a control system of quality of filling of melted metal successfully is carried out by identification of spectral characteristics of objects at their decomposition and is offered by authors linearly—a harmonious row.

References

1. Arustamyan AI, Minakov VF (2006) Device for control of provision of a ladle with the melted metal. The patent for the useful model, RUS 69233. 25 April 2006 (in Russian)
2. Arustamyan AI, Minakov VF, Minakova TE (2006) Automation of control of provision of a ladle when pouring cast iron. In: Materials XXXV of scientific and technical conference on results of work of the faculty, graduate students and students of NCSTU, vol 1. Stavropol: NCSTU, pp 55–56 (in Russian)
3. Arustamyan AI, Minakov VF (2006) Synthesis of a spectral characteristic of sensitivity of sensors. Materials X of the regional scientific and technical conference "High School Science —to the North Caucasian Region". Stavropol: NCSTU, pp 100–101 (in Russian)
4. Astel A, Małek S (2008) Multivariate modeling and exploration of environmental n-way data from bulk precipitation quality control. J Chemometr 22(11–12):738–746
5. Fujiwara K, Kano M, Hasebe S, Takinami A (2009) Soft-sensor development using correlation-based just-in-time modeling. AIChE J 55(7):1754–1765
6. Jämsä-Jounela SL (2001) Current status and future trends in the automation of mineral and metal processing. Control Eng Pract T9(9):1021–1035

7. Kano M, Nakagawa Y (2008) Data-based process monitoring, process control, and quality improvement: recent developments and applications in steel industry. Comput Chem Eng 32:12–24
8. Maslov VI, Arustamyan AI, Minakov VF (2012) Optical control in a control system of quality of production of piston rings. Scientific and technical messenger of information technologies, mechanics and optics. 2(78):16–20 (in Russian)
9. Maslov VI, Arustamyan AI (2012) Improvement of quality of production of piston rings. Scientific and technical sheets of the St. Petersburg state polytechnical university. Sci Educ 1(142):132–136 (in Russian)
10. Maslov VI, Arustamyan AI, Minakov VF (2012) Mathematical model of spectral characteristics of radiations and sensors. In: Materials of the 12th international scientific and methodical conference "Informatics: problems, methodology, technologies", The Voronezh state university, on February 8–10, 2012 Voronezh: publishing and printing center of the Voronezh state university. pp 241–242 (in Russian)
11. Maslov VI, Arustamyan AI, Minakov VF (2013) Remote control in a control system of quality of filling of metal. Mod Eng: Sci Educ 3:450–459 (in Russian)
12. Maslov VI, Arustamyan AI, Minakov VF (2012) WEB services in a control system of quality machine-building продукции. Mod Mech Eng Sci Educ 2:472–478 (in Russian)
13. Maslov VI, Minakov VF (2012) E elasticity of quality at the price and expenses. Stan Qual 9(903):88–90 (in Russian)
14. Maslov VI, Minakov VF (2012) Kritery of efficiency in a control system of quality of production of the enterprise. Scientific and technical sheets of the St. Petersburg state polytechnical university. Inform Telecommun Manage 6(162):179–184 (in Russian)
15. Mikkelsen Ø, Skogvold SM, Schrøder KH (2005) Continouos heavy metal monitoring system for application in river and seawater. Electroanalysis 17(5–6):431–439
16. Minakov VF, Arustamyan AI (2010) Technology of control of provision of a ladle when pouring metal. Mater Sci Questions 4(64):72–78 (in Russian)
17. Minakov VF, Makarchuk TA, Shchugoreva VA (2014) The WEB technologies 2.0 in a control system of quality. Res J Int Stud 1–1(20):70–72 (in Russian)
18. Minakov VF, Minakova TE (2014) Linearly—the harmonious analysis of optical spectra. In: Materials of the 3rd scientific and practical internet-conference interdisciplinary researches in the field of mathematical modeling and informatics. Ulyanovsk. pp 288–294 (in Russian)
19. Pelleg J (2013) Mechanical properties of materials. Series: solid mechanics and its applications, vol 190. Springer, Berlin, 634 p
20. Wang H et al (2009) Data driven fault diagnosis and fault tolerant control: some advances and possible new directions. Acta Automatica Sin 35(6):739–747

Chapter 10
Acoustic Emission Monitoring of Leaks

Evgeny Nefedyev

Abstract Our research is devoted to the study of the possibility of determining the parameters of leaks in the pipelines of ITER by the acoustic emission method. Research results are presented for leaks in pipelines, length from 900 mm to 150 m. Flow was carried out with steam and water through fatigue cracks in size from 5 to 39 mm in the pipelines. We investigated regularities of changes of the amplitude, temporal and spectral parameters of acoustic emission signals. The loss through flow and the distance from the source of leaks to the AE sensors were changed. Methodology was developed, which allows us to determine the coordinates and parameters of leakage through information on the AE signals.

Keywords Acoustic emission · Leaking · A pipeline · A crack · Flow rate · Spectral parameters · Coordinates of leak

Introduction

Acoustic emission (AE)—modern NDT method [1–3]. This method is widely used for inspection of pressure vessels [4–6] and pipelines [7]. The main source of AE signals are cracks under static and cyclic fracture [8, 9]. Leaks in pipelines is a dangerous source of AE signals also [10]. The main tools of the study of AE from leaks are the methods of spectral analysis [11, 12].

The study of remote positioning of leaks in the pipelines of the cooling system ITER was carried out. The aim of the research was development of existing recommendations. Various experiments have been made in the study of leaks on the program of ITER by means of the method of acoustic emission (AE). Objects in the form of a pipe length from 900 mm to 150 m have been studied. There have been many experiments, including:

E. Nefedyev (✉)
Central Boiler and Turbine Institute (CKTI), Saint-Petersburg, Russia
e-mail: ne246@ya.ru

© Springer International Publishing Switzerland 2015
A. Evgrafov (ed.), *Advances in Mechanical Engineering*, Lecture Notes in Mechanical Engineering, DOI 10.1007/978-3-319-15684-2_10

- Study of the amplitudes AE signals at various flow rates of liquid in a pipe length of 150 m.
- Study of the spectrum of AE signals at different lengths of fatigue cracks.
- Study of changes in the spectral composition of AE signals from leakage at different distances from the source.
- Obtaining the regression dependencies between parameters of fluid flow, the distance to the leak, amplitude and frequency characteristics of AE signals.

Some of the results of these experiments are presented next.

Part 1

The study of the amplitudes of AE signals was made at various flow rates of liquid in a pipe length of 150 m. For AE research, control of the leakage of water through the crack in the pipeline is made. The installation scheme is shown in Fig. 10.1. In the central part of the pipe (1), there is a welded-in section of the pipeline (2) $\varnothing 101.6 \times 5.7$ mm wall thickness (of steel 08CR18NI10TI) with a crack. After filling the pipe with water, internal pressure was created with the aid of a nitrogen cylinder (3). The amount of air pressure is maintained at a given level by means of a pressure regulator (4). The water coming out of the cracks was collected in a container with the purpose of measuring the flow rate. For registration of the AE we

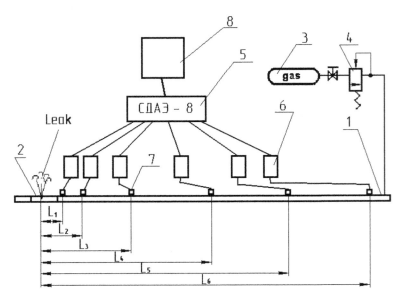

Fig. 10.1 Experimental setup. _1_ tube $\varnothing 110 \times 10$ mm, a length of 150 m; _2_ the pipeline section with crack, $\varnothing 101.6 \times 5.7$ mm; _3_ cylinder with nitrogen; _4_ pressure regulator; _5_ AE-system; _6_ preamplifier; _7_ AE-sensor; _8_ computer

The dependence of the amplitude of AE signals from the distance to leak
at a cost: line 1 -0.23л/m, line 2 -0.35л/m, line 3 -0.6л/m, line 4 -2n/m.

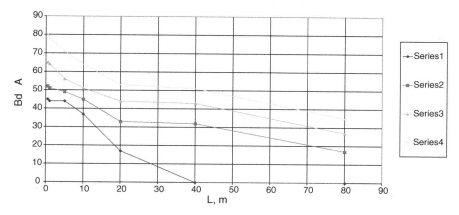

Fig. 10.2 The dependence of the amplitude of AE signals (in dB) at a distance from the leak in various flow rates: line 1—0.23 l/m, line 2—0.35 l/m, line 3—0.6 l/m, line 4—2 l/m

used a system of SDAE-8 (5) with sensors AP-206 (7). The used frequency range was in the range of 20 Hz–150 kHz. Measurement of amplitude of AE signals was conducted with different pressures in the pipe and various water leaks through the crack. These data are summarized and shown in Fig. 10.2.

Conclusions of section 1

1.1. At the minimum leak rate of water in this experiment (0.23 l/min), the leak can be securely detected by the AE method at distances from 0 to 20 m.

1.2. Leakage rate of less than 0.23 l/min can be detected by putting an AE sensor at the minimal distance from the crack. The minimum leak rate that can be detected in such conditions depends on the level of noise generated by the working equipment.

1.3. At higher leak rates (0.35–2 l/min), detection distance can be more than 80 m using the same sensitivity level as in 1.1.

Part 2

Study of spectral composition of AE signals at different length of a fatigue crack was conducted. The research data was made on a model pipeline Ø76.3 × 5.2 mm, a length of 900 mm. On the pipe, at a distance of 40, 430 mm from the cracks, two waveguides were installed with AE sensors to get AE signals radiated by a leak. The sensors recorded continuous signals from water leaks. The spectral composition of the AE signals are shown in Figs. 10.3, 10.4 and 10.5.

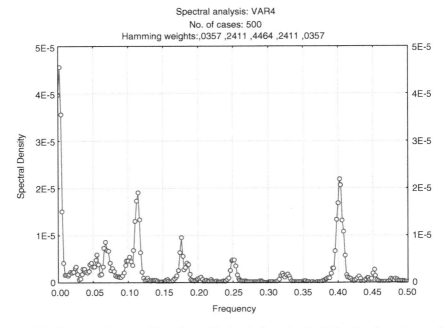

Fig. 10.3 The spectral composition of the AE signals from the liquid flowing through a crack length of 5 mm (scale on the X-axis MHz; the scale on the Y-axis In/MHz)

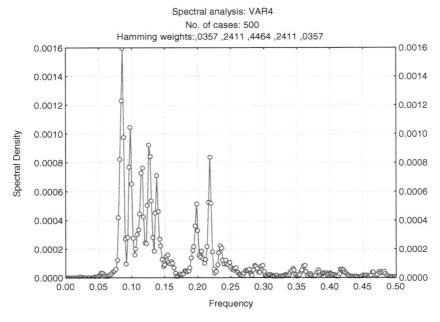

Fig. 10.4 The spectral composition of the AE signals from the liquid flowing through the crack length of 12 mm (scale on the X-axis MHz; the scale on the Y-axis In/MHz)

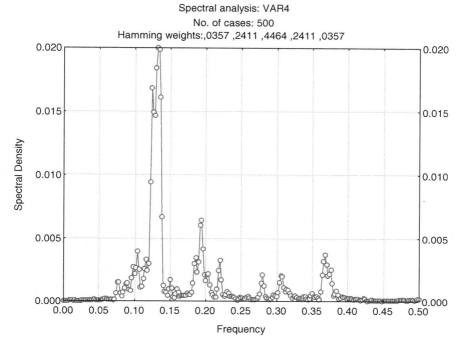

Fig. 10.5 The spectral composition of the AE signals from the liquid flowing through the crack length of 39 mm (scale on the X-axis 1 MHz; the scale on the Y-axis In/MHz)

Conclusions of section 2

2.1. The amplitude of AE signals increases with the increase of the sizes of cracks and fluid flow;
2.2. The frequency of AE signals is reduced with the increase in the length of the crack and fluid flow.
2.3. The spectral composition of the signals are heterogeneous from the small size of the model pipeline and non-stationary.
2.4. It should be noted that signal spectrum is significantly different at different locations of the AE sensor. This apparently is connected with the resonance phenomena of the limited size of the model pipeline. Presumably the difference in the signals may be due to the change of the quantity of water in a small pipe.

Part 3

The change in the spectral composition of AE signals from leakage at a different distance from the source was studied next. One of the directions of further research was suggested to change the size of the model by increasing its length to the size of real pipelines, which would allow us to exclude the above-mentioned shortcomings. This phase of the research was carried out on a section of the pipeline Ø106 × 9 mm, with length of 150 m (Fig. 10.6). In the middle part of the pipeline was welded-in pipe section with a defect in the form of crack. The pipe was completely filled with water. Tests were carried out at room temperature. Under the action of constant internal pressure, there was a discharge of coolant through the end-to-end crack (Fig. 10.7). At the same time on the pipe, at a distance of 500 mm, 1, 5 and 10 m were installed broadband AE sensors (with a bandwidth of 10 kHz– 1 MHz) to detect AE signals radiated by a leak. Continuous signals from leaks were recorded using a broadband digital oscilloscope. The signals were sent to a computer, where they were evaluated for their spectral composition. The spectral composition of the AE signals at different distances from the leak is shown in Figs. 10.8 through 10.11.

From the analysis of spectral composition of the signal, it can be seen that at a distance of 0.5 m a spectral component of ∼0.37 MHz is dominant and a minor low-frequency (0.16 MHz) peak is present (Fig. 10.8). At a distance of 1 m from leaks, AE signals have several high frequency peaks at 0.29–0.4 MHz. In the low-frequency part of the spectrum, the peak appears at 0.08 MHz (Fig. 10.9).

At a distance of 5 m from leaks, the maximum peak frequency was slightly lower at 0.29 MHz and the low-frequency peaks were over 0.05–0.1 MHz. However, in this case the energy of the low-frequency peaks became stronger, getting closer to that of high frequency peaks (Fig. 10.10). At a distance of 10 m from the leaks, we see that almost all of the energy of AE signals due to leaks is in the low-frequency peak at 0.08 MHz (Fig. 10.11).

Fig. 10.6 General view of the 150-m pipeline in the left part of the stand

Fig. 10.7 Type of leaks from the 150-m pipeline

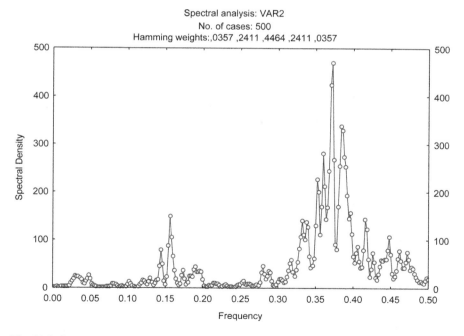

Fig. 10.8 Range of signals at a distance of 0.5 m from leaks

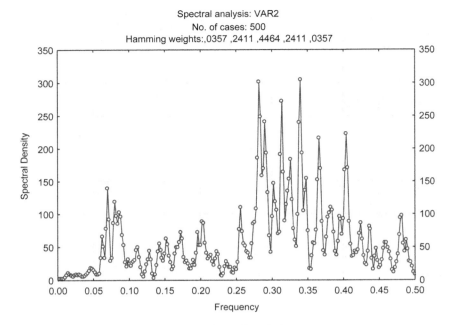

Fig. 10.9 Range of signals at a distance of 1.0 m from leaks

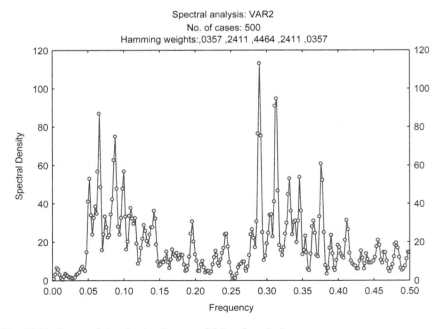

Fig. 10.10 Range of signals at a distance of 5.0 m from leaks

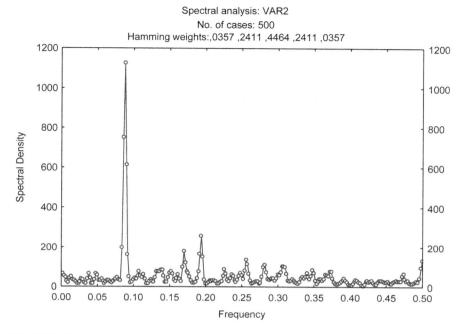

Fig. 10.11 Spectrum of the signals on the distance 10.0 m from leaks

Conclusions in section 3

These results show that the spectral distribution of AE signals from the leak changes with the distance to the sensor. At small distances, the main energy is concentrated in the high-frequency range of 300–400 kHz. As we move through the pipe away from the leak source, the low-frequency component of the spectrum of the signal starts to dominate with the main frequency of 80 kHz.

Part 4

We attempt to get regression dependencies between the parameters of fluid flow through the crack, the distance to the sensor from leak and the amplitude and frequency characteristics of AE signals. Treating methods of mathematical statistics data from Part 1 on the dependence of the amplitude of AE signals on the distance to leak and flow of the liquid through a crack, you can get the regression equation relating these three parameters. The expression for the amplitude of the AE signal is

$$A = f(X, R) = C0 + C1 \ X + C2^*R, \tag{10.1}$$

where A is the amplitude of the AE signals in decibels, X is the distance from the AE sensor to leak in m, R is fluid flow through the crack in l/min, C0, C1 and C2 are constants. In this case,

$$C0 = 49.5 \text{ dB}, C1 = 0.46 \text{ dB/m}, C2 = 11.6 \text{ dB}^*\text{min/l}.$$

In addition based on the laws of acoustics, and the results of Part 1, one can derive the expression for the coordinates of the leak by the ratio of the amplitudes of AE signals, detected at two AE sensors, regardless of the flow through the crack.

$$X = X1 + 0.5^*[X12 - (A1 - A2)/K] \tag{10.2}$$

where X is the coordinate of the leak, X1—coordinate of the first AE sensor, X12—the distance between the first and second AE sensors, (A1–A2)—the difference of the amplitudes of AE signals, detected by the first and second AE sensors, K—the coefficient of attenuation in dB/m.

Data analysis allows us to calculate the coordinates of a defect or leakage, knowing delay τ and the speed of propagation of acoustic waves in the pipeline, V.

$$X = X1 + 0.5^*[X12 - \tau^*V] \tag{10.3}$$

Thus, the data obtained in the preliminary experiments and formulas (10.1)–(10.3) allow one to determine where the coordinates of the leaks, are and their leak rate.

Conclusions

The results of these studies show that, on the basis of data of analog and digital measurements it is possible to develop a methodology for identifying leaks using the AE method, as well as determining its coordinates and leak rate.

References

1. Greshnikov VA, Drobot JB (1976) Acoustic emission. Standards Publishing House, p 272
2. Rules of the acoustic emission method application during inspection of boilers, vessels, apparatus and process piping. The Safety Regulations 03-593-03. M. 2003
3. Ivanov VI, Belov IE (2005) The method of acoustic emission. In: Klyuev VV (ed) vol 7. Machine-Building, 825 pp
4. Gomera VP, Sokolov VL, Fedorov VP (2008) Implementation of acoustic emission method to the conventional NDT structure in oil refinery. J Acoust Emiss (Acoustic Emission Group, Ehcino, CA, USA) 26:279–289

5. Kabanov VS, Sokolov VL, Homer VP, Fedorov VP (2011) The experience of application of acoustic emission technique for the inspection of pressure vessels. Chem Eng 4:12–15
6. Catty J (2009) Acoustic emission testing—defining a new standard of acoustic emission testing for pressure vessels. Part 1: Quantitative and comparative performance analysis of zonal location and triangulation methods. J Acoust Emiss 27:299–313
7. Kovalev DN, Nefedyev EJ, Tkachev VG (2012) Acoustic emission control testing of steel corrugated pipes of circular and static loading. In: Radkevich MM, Evgrafova AN (eds) Modern engineering, science and education: materials of the 2nd international nauch-practical use conference. SPb Publishing House Polytechnic University, pp 382–390
8. Nefedyev EJ et al (1986) Connection sizes of microcracks with the acoustic emission parameters and structure of deformed steel rotor. Defectoscopy 3:41–44
9. Nefedyev EJ et al (1987) The control of the fatigue cracks growth in cast steel method of acoustic emission. Probl Strength 1:28–31
10. Nefedyev EJ (2013) The use of acoustic emission method with spectral analysis of signals to determine the parameters of a leak in the pipeline ITER. In: Radkevich MM, Evgrafova AN (2013) Modern engineering, science and education: materials of the 3rd international nauch-practical use conference. SPb Publishing house Polytechnic University, pp 347–355
11. Otnes RK, Enochson L (1978) Applied time series analysis. A Wiley-Interscience Publication, New York
12. Bendat JS, Piersol AG (1980) Engineering applications of correlation and spectral analysis. A Wiley-Interscience Publication, New York

Chapter 11
Dynamic Model of the Two-Mass Active Vibroprotective System

Mikhail J. Platovskikh

Abstract While noise and vibration reduction requirements for modern power installations have become tougher and tougher, possibilities of passive vibration insulation are almost settled. Thus active vibroprotective systems (AVPS) have become crucial. In this paper we investigate the dynamics of two-stage AVPS and define the areas of efficiency and stability.

Keywords Active vibroprotective system · Coefficient of active vibration insulation · Electrodynamic vibrator · Hurvitz stability criterion

Introduction

The most effective way to isolate vibrations is to reduce variable forces at their sources and in energy transmissions (internal combustion engines, cog gearings, electric motors, etc.). But at the design stage the most important task is to fulfill the main functional objective—provide efficient energy transmission from the source to a receiver. At the same time the design has to fulfill strength and resource characteristics. Vibroactivity often gets second priority at this point. Thus this method of vibration reduction has limited perspective.

A large number of vibroprotective systems (VPS) are based on the use of a variety of shock-absorbers developed for protection of technical and biological objects against vibrating excitement at low frequencies. Such VPS are known as passive. However they are ineffective in many applications, for example, in case of vibration spectrum changes in time [1].

Automated vibroprotective systems called active (AVPS) [2, 3] have recently been found to be useful. Development of effective active vibration insulation systems for low-frequency vibration of various mechanisms has been the aim of the

M.J. Platovskikh (✉)
University of Mines, St. Petersburg, Russia
e-mail: mplat.63@gmail.com

© Springer International Publishing Switzerland 2015
A. Evgrafov (ed.), *Advances in Mechanical Engineering*, Lecture Notes in Mechanical Engineering, DOI 10.1007/978-3-319-15684-2_11

work of many researchers over the last few decades. Generally such systems control can be built on a principle of disturbance compensation, compensation of a deviation of adjustable size, or on a combination of both methods.

Our experience of active vibroprotective systems creation shows that the most perspective is electrodynamic VPS based on electrodynamic vibrator [4] due to its crucial benefits: completeness of reproduction of variable efforts, comparative ease of implementation and management, low sensitivity to negative factors of environment.

A characteristic feature of active systems is the use of active circuits, consisting of measuring, amplifying and actuator components. Actuators form the force that can reduce the influence of dynamic loads on the protected object.

In active VPS information on the nature of a disturbance, its spectrum and amplitude structure is necessary to control it (management). The role of sources of this information is performed by electric converters of vibrations acting here as parameters converters of movement (force, acceleration, motion) to electric signals (voltage, current). The converters (motion, force, accelerometers and other sensors) should have rather wide frequency ranges (at least, five times wider than the frequency range of the measured signal) and small nonlinear distortions. Electric control signals should be proportional to disturbing force $Q(t)$. If frequency and amplitude of external influence change, frequency and amplitude of control current (voltage) should change in a similar manner.

In works [5–8] the electromagnetic actuator was used as an active element and the air spring acted as the passive element. It was demonstrated that this hybrid control system could provide a better isolation performance than the passive system alone. Gardonio studied the theoretical effectiveness of various control strategies for active vibration isolation. In their work [5, 6], the minimization of the total power transmission through the mounts to the receiver was compared with several control strategies: the minimization of the axial velocities or forces.

An oscillatory system with an electrodynamics vibrator can be considered as an object of automatic control that can be applied to research developed in the theory of automatic control methods [2–4, 9–11]. In order to formulate a system of equations describing the dynamics of a mechanical oscillatory system with the electrodynamics generator of force, it is necessary to consider both mechanical movement, and electric processes in a conductor chain (the mobile coil). In electrodynamics devices the current for creation of force results from movement of the conductor or its points of suspension. The principle of management on an absolute deviation and disturbance is considered in work [1, 12–20].

Model of the Active Vibroprotective System

Let's investigate an AVPS model with various schemes of connection of the electrodynamics vibrator and the appendix of external dynamics loading. Thus in all cases we pose the problem of decrease in dynamic load of the base. Modeling is made in the field of low frequencies.

Fig. 11.1 Model of two-stage
AVPS

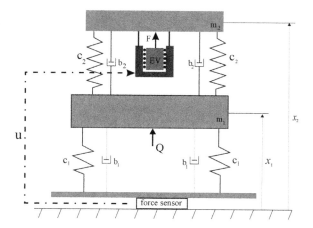

Let's consider the AVPS model represented in Fig. 11.1 and assume that the problem of an active vibration insulation of some elastic strengthening mass (m1) in near to the resonance range (rather low frequencies are considered) The external disturbing force $Q(t)$ operates on mass. For installation of the vibrator the additional mass (m$_2$) fastened to the isolated weight by means of elastic elements (c$_2$) is introduced. Parallel to elastic links also dissipative elements with factors of resistance b_1, b_2 are included. Between an intermediate plate and the motionless basis the sensor of force D, reformation of the force operating on a plate in an operating signal (tension of u on coil clips) is established.

The system is described by the equations:

$$m_1 \ddot{x}_1 = -c_1 x_1 + c_2(x_2 - x_1) - b_1 \dot{x}_1 + Q - F,$$
$$m_2 \ddot{x}_2 = -c_2(x_2 - x_1) - b_2 \dot{x}_2 + F, \tag{11.1}$$
$$L\frac{di}{dt} + Ri = U - B\ell(\dot{x}_2 - \dot{x}_1).$$

The first two equations of system (11.1) characterize movement of mass of *m1* and *m2*, the third—electrodynamics balance in a chain of the mobile coil of the vibrator. In these equations *x1* and *x2*-absolute coordinates of masses; $Q = Q(t)$—external revolting force; *i*—current in a chain of a winding of management of the electrodynamics vibrator; $F = F(i)$—ponderomotive force depending on a current in a winding; *L, R*—inductance and active resistance of a winding of management; *U*–electric tension on a winding of the mobile coil. According to the Ampere law—$F(i) = B\ell i$. For creation of necessary electric tension on a winding of management of an electromagnet, between the sensor of force and an electromagnet the amplifier which, in turn, can contain a correcting chain for providing dynamic requirements to VPS (operating speed, accuracy of working off of a setting signal, etc.) is established.

In the considered system it is supposed that time constants of various elements in a chain of feedback less time constant of the T_e vibrator is essential.

Management is entered as negative feedback on the total force operating on the basis (sensor D): $U = -k_U(b_1\dot{x} + c_1 x_1)$ where factor of proportionality between tension on a winding and force $k_U = k_d\,k_{am} > 0$ (k_f, k_{am}—factors of sensitivity of the sensor of force and strengthening of the amplifier).

Having passed to new parameters:

$$\lambda_1^2 = \frac{c_1}{m_1}, \quad \lambda_2^2 = \frac{c_2}{m_2}, \quad \lambda_{12}^2 = \frac{c_2}{m_1}, \quad \beta_1 = \frac{b_1}{m_1}, \quad \beta_2 = \frac{b_2}{m_2},$$
$$T = \frac{L}{R}, \quad \rho = \frac{B\ell}{m_2}, \quad k_1 = \frac{k_u b_1}{R}, \quad k_2 = \frac{k_u c_1}{R}, \quad \mu = \frac{B\ell}{R}.$$

and having applied to (11.1) Laplace transformation with parameter p ($x \Rightarrow \overline{x}$, $i \Rightarrow \overline{i}\,\overline{q} = \frac{1}{m_1}\overline{Q}$), we will receive a linear system for definition $\overline{x_1}$, $\overline{x_2}$, $\overline{i}$:

$$\left(p^2 + p\,\beta_1 + \lambda_1^2 + \lambda_{12}^2\right)\overline{x_1} - \lambda_{12}^2 \overline{x_2} + \rho_1\,\overline{i} = \overline{q},$$
$$- \lambda_2^2\,\overline{x_1} + \left(p^2 + p\,\beta_2 + \lambda_2^2\right)\overline{x_2} - \rho_2\,\overline{i} = 0, \qquad (11.2)$$
$$\left(p\,(k_1 - \mu) + k_2\right)\overline{x_1} + p\,\mu\overline{x_2} + (1 + T_э)\,\overline{i} = 0.$$

Numerically solving system (11.2), (the programming environment of MAT-LAB is used), we find $\overline{x_1}$, $\overline{x_2}$, $\overline{i}$. Laplace's transform at most, transferred to the fundament (through an elastic element c_1 and b_1 damper): $\overline{R_f} = (b_1\,p + c_1)\,\overline{x_1}$.

Let's consider the passive system (Fig. 11.2) corresponding to the two-mass system we studied.

The movement equation corresponding to it is

$$m_1\ddot{x}_1 = -c_1 x_1 - b_1\dot{x}_1 + Q. \qquad (11.3)$$

Carrying out Laplace transformation, we will receive

$$\overline{x_1} = \frac{\overline{q}}{\left(p^2 + p\,\beta_1 + \lambda_1^2 + \lambda_{12}^2\right)}.$$

Fig. 11.2 Scheme of passive VPS

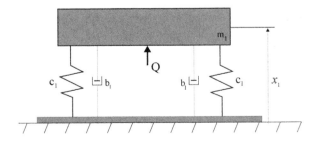

Laplace transformation of the force transferred to the basis at a passive vibration insulation:

$$\overline{R_{f\,0}} = (b_1 p + c_1)\overline{x_1} = \frac{(b_1 p + c_1)}{\left(p^2 + p\,\beta_1 + \lambda_1^2 + \lambda_{12}^2\right)}\,\overline{q}. \qquad (11.4)$$

The indicator defining efficiency of application of an active system in relation to the passive—factor of an active vibration insulation γ:

$$\gamma = \frac{\overline{R_\phi}(p)}{\overline{R_{\phi\,0}}(p)}.$$

Condition of an effective active vibration insulation is the inequality $\gamma > 1$. Substituting $p = i \cdot \omega\,(i = \sqrt{-1})$, ($\omega$—frequency of oscillation), we will receive frequency dependence for the vibration insulation coefficient. On Fig. 11.3 we will receive frequency dependence for vibration insulation coefficient $\mathrm{mod}(\gamma(\omega))$ depending on frequency at the following values of parameters: $m_1 = 100$ kg, $m_2 = 10 \div 50$ kg, $c_1 = 5 \times 10^3$ N/m (natural frequency of passive system c_1–m_1 equal $\lambda_1 \approx 14\ c^{-1}$), $c_2 = 1 \times 10^2$ N/m, $R = 40\ \Omega$, $L = 5$ mGn, $B\,\ell = 100\ \mathrm{T_1 \cdot m}$, $b_1 = b_2 = 70$ Nc m, $k_U = 3$.

Fig. 11.3 Curves dependences of factor γ on frequency ω, constructed for various values of mass m_2: curve 1—$m_2 = 10$ kg, 2—20 kg, 3—30 kg, 4—40 kg, 5—50 kg

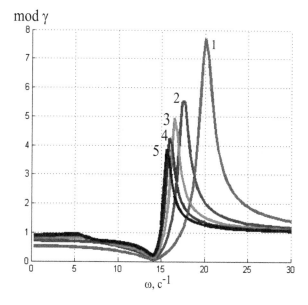

Stability of the system

We investigate stability of this system, applying Gurvits's criterion. The characteristic equation of system (11.2) will look like:

$$\Delta(p) = \begin{vmatrix} p^2 + \beta_1 p + \lambda_1^2 + \lambda_2^2 & -\lambda_{12}^2 & \rho_1 \\ -\lambda_2^2 & p^2 + \beta_2 p + \lambda_2^2 & -\rho_2 \\ (k_1\mu)p + k_2 & \mu p & T_{\ni}p + 1 \end{vmatrix} = 0. \qquad (11.5)$$

Having opened the determinant, we will receive an equation of the fifth degree on p:

$$a_0 p^5 + a_1 p^4 + a_2 p^3 + a_3 p^2 + a_4 p + a_5 = 0,$$

where factors *of a_1 ... a_2*:

$$
\begin{aligned}
a_0 &= T_e, \\
a_1 &= T_e(\beta_1 + \beta_2) + 1, \\
a_2 &= T_e\left(\lambda_1^2 + 2 \cdot \lambda_2^2 + \beta_1\beta_2\right) - \rho_1(k_1 - \mu) + \rho_2\mu + \beta_1 + \beta_2, \\
a_3 &= T_e \beta_2 \left(\lambda_1^2 + \lambda_2^2\right) + T_e \beta_1 \lambda_2^2 - \rho_1(k_1 - \mu)\beta_2 - \rho_1 k_2 + \rho_2\beta_1\mu \\
&\quad + \lambda_1^2 + 2 \cdot \lambda_2^2 + \beta_1\beta_2, \\
a_4 &= T_e \lambda_2^2 \left(\lambda_1^2 + \lambda_2^2\right) - \rho_1\lambda_2^2\mu - \rho_1(k_1 - \mu)\lambda_2^2 - \rho_1\beta_2 k_2 \\
&\quad + (\rho_2\mu + \beta_2)\left(\lambda_1^2 + \lambda_2^2\right)\mu - T_e \lambda_2^2\lambda_{12}^2 + \beta_1\lambda_2^2 + \rho_2(k_1 - \mu)\lambda_{12}^2, \\
a_5 &= \rho_2\lambda_{12}^2 k_2 - \rho_1\lambda_2^2 k_2 + \lambda_2^2\left(\lambda_1^2 + \lambda_2^2\right) - \lambda_1^2\lambda_2^2.
\end{aligned}
\qquad (11.6)
$$

Fig. 11.4 Stability ranges of system (are located between an *axis* of abscissae and the corresponding *top border*) at various values of coefficient of amplification k_U. Values of other parameters the same, as in Fig. 11.3. At $k_U > 4.2$ system is unstable

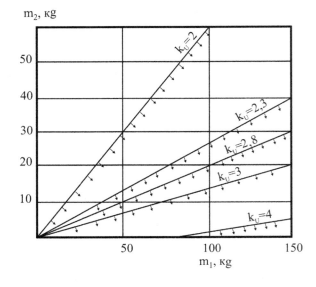

Gurvits's criterion and this case [14] is reduced to the following inequalities:

$$a_i > 0, \, i = 0, \ldots, 5;$$
$$\Delta_2 = a_1 a_2 - a_0 a_3 > 0; \tag{11.7}$$
$$\Delta_4 = (a_1 a_2 - a_0 a_3) \cdot (a_3 a_4 - a_2 a_5) - (a_1 a_4 - a_0 a_5)^2 > 0.$$

Here Δ_2, Δ_4—Gurvits's determinants of orders 2 and 4 are corresponding characteristic equations of the fifth degree. Stability ranges numerically calculated by formulas (11.6) and (11.7) are shown on Fig. 11.4.

Conclusion

From the conducted calculations it is visible that AVPS corresponding to the considered model is effective (up to $\gamma \approx 8$) also it exhibits stability in a range of frequencies of $15 \div 25 \, c^{-1}$. Expansion of this range is possible by increase factor of back coupling k_U, however thus the system loses stability. The increase in a magnetic stream B allows us to expand a frequency range of efficiency, however efficiency size is considerably reduced. Thus, more effective functioning of AVPS constructed on the basis of the considered model needs introduction in a control system of the additional correcting links based on use of programming controllers.

These results can be used in designing vibroprotective systems in the conditions of impact of broadband non-stationary vibration: seats of drivers of motor transport, workplaces of operators of mining combines and pilots of planes.

References

1. Harris CM, Piersol AG (2002) Harris' shock and vibration handbook, (5th edn). McGRAW-HILL
2. Frolov K, Fuhrman F (1980) Applied theory of vibroprotective systems. Machinery construction (Mashinostroyeniye) (in Russian)
3. Kolovsky M (1976) Automatic control of vibroprotective systems. Science (Nauka) (in Russian)
4. Genkin M, Yablonsky V (1975) Electrodynamics vibrators. Machinery construction (Mashinostroyeniye) (in Russian)
5. Gardonio P, Elliott SJ (1999) A study of control strategies for the reduction of structural vibration transmission. J Vib Acoust 121:482–487
6. Gardonio P, Elliott SJ, Pinnington RJ (1997) Active isolation of structural vibration on a multiple-degree-of freedom system. Part I: the dynamics of the system. J Sound Vib 207 (1):61–93
7. Gardonio P, Elliott SJ, Pinnington RJ (1997) Active isolation of structural vibration on a multiple-degree-of freedom system. Part II: effectiveness of active control strategies. J Sound Vib 207(1):95–121

8. Lee JW, Li X, Cazzolato BS, Hansen CH (2001) Active vibration control to reduce the low frequency vibration transmission through an existing passive isolation system. In: Proceedings of ICSV8, Hong Kong, pp 325–332

9. Genkin M, Elezov V, Yablonsky V, Friedman F (1975) Development of methods of a vibroprotection, volume. Methods of a vibration isolation of cars and the attached designs. Science (Nauka) (in Russian)

10. Vuong NV, Zasyadko A (2011) Estimation of frequency range of controlled vibrodamping efficiency. National Research Irkutsk State Technical University. Cybernetics, information systems and technologies, № 4, 2011 (in Russian)

11. Yeliseyev S, Zasyadko A (2004) Vibroprotection and vibration insulation as management of fluctuations of objects. Modern technologies. System analysis. Modeling. Irkutsk: Publishing house of IRGUPS. № 1. p. 26–34 (in Russian)

12. Kaczmarek K, Novak A (2007) Optimization of the vibroisolation systems of crane cabin using genetic algorithm. Institute of Mathematic, Silesian Technical University, Gliwice

13. Li X, Cazzolato BS, Hansen CH (2002) Active vibration isolation of diesel engines in ships. In: Ninth international congress on sound and vibration. Department of Mechanical Engineering, Adelaide University

14. Genkin M, Elezov V, Yablonsky V (1971) Methods of active clearing of vibrations of mechanisms, volume. Dynamics and acoustics of cars. Science (Nauka) (in Russian)

15. Gardonio P, Elliott SJ (2000) Passive and active isolation of structural vibration transmission between two plates connected by a set of mounts. J Sound Vib 237(3):483–511

16. Frolov K, Goncharevich I, Likhnov P (1996) The Infrasound, the vibration, the person. Machinery construction (Mashinostroyeniye). (in Russian)

17. Calcaterra, SD (1969) Active vibration isolation of human subjects from severe dynamic environments. Papers of American Society Mechanical Engineers, Vibrations, № 65

18. Shimanov V (1974) Tool of active vibroprotective. Science (Nauka), (in Russian)

19. Genkin M, Yablonsky V (1977) Active vibroprotective systems. Vol. Anti-vibration systems in machines and mechanisms. Science (Nauka), (in Russian)

20. Bogko A, Gal A, Gurov A (1988) Passive and active vibroprotection of ship mechanisms. Vol. Passive and active vibroprotection of ship mechanisms. Shipbuilding (Sudostroenie), Leningrad, (in Russian)

Chapter 12
Structural and Phase Transformation in Material of Blades of Steam Turbines from Titanium Alloy After Technological Treatment

Margarita A. Skotnikova, Galina V. Tsvetkova,
Aleksandra A. Lanina, Nikolay A. Krylov and Galina V. Ivanova

Abstract By using optical metallography, electron microscopy, ray analysis and microspectral analysis, we devised a systematic research method to explore structural and phase transformation in blades material of steam turbines from a titanium alloy VT6 after technological treatments on different modes.

Keywords Steam turbines · Titanium alloy · Structural and phase transformation · Electron microscopy

Introduction

In power mechanical engineering the problem of erosion destruction of steam turbines blades is the most important. The difficulty of solving this problem is that, until recently, it was impossible to find the general relationship between wear and a structural—phase state of the surface and axial layers of blades material.

There is an opinion [1], that the greatest resistance to effect of exposure steam drops should be two-phase alloys with low internal stresses and high plastic

M.A. Skotnikova (✉) · G.V. Tsvetkova · A.A. Lanina · N.A. Krylov · G.V. Ivanova
St. Petersburg State Polytechnic University, Saint Petersburg, Russia
e-mail: skotnikova@mail.ru

G.V. Tsvetkova
e-mail: tsvetkova_gv@mail.ru

A.A. Lanina
e-mail: lanina-alexandra@yandex.ru

N.A. Krylov
e-mail: cry_off@mail.ru

G.V. Ivanova
e-mail: galura@yandex.ru

© Springer International Publishing Switzerland 2015
A. Evgrafov (ed.), *Advances in Mechanical Engineering*, Lecture Notes
in Mechanical Engineering, DOI 10.1007/978-3-319-15684-2_12

properties, during which time of stress there exist phase transformations strengthening the subject material. At effect of exposure steam drops, field sizes of a stressing are commensurable with sizes of structural components, and redistribution of internal stresses between them is impossible. The role of individual durability, intensity, a chemical composition and phase transformations in separate structural components therefore increases.

Two-phase (α+β)—titanium alloys have found wide application in turbine-structure. As follows from earlier carried out works [2–6], on the basis of complex research of mechanisms of formation and decomposition nonequilibrium β(α)- and α(β)-phases, redistribution between them of alloying elements, physical and mechanical properties of the deformed titanium alloys of a different alloying, and also the established laws, the generalized kinetic diagram, Fig. 12.1, has been constructed.

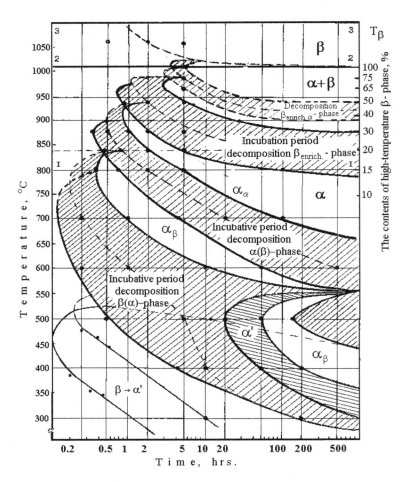

Fig. 12.1 Generalized kinetic diagram of structure and phase transformation in titanium alloys

Thus we took into account not only a temperature of heating absolute (T), but also produced the identical contents of high-temperature β-phases (T_β). We established that the more nonequilibrum $\beta(\alpha)$- and $\alpha(\beta)$-solid solutions contained the same alloying elements; their polystage decomposition occurred at lower temperatures, and for greater time. It is shown that titanium blanks possess high technological properties in a temperature-time interval two-phase ($\alpha+\beta$)-region at temperatures identical (50 %) contents high-temperature α- and β-phases (T_{50}) and near to a temperature of transition in single-phase β-region (T_{75}).

The product possesses material with high operational properties after an iso-thermal exposure near to temperature T_{15}. In these cases α- and β-phase compo-nents appear enriched, as α- and β-stabilizing alloying elements, in these processes of a polygonization that causes increase of plasticity characteristics and allows crack origin of alloys to develop.

Thus, construction of serial curves of increase of the contents high-temperature β-phases depending on a temperature of heating (a method of hardenings), will allow us to estimate the resulting temperatures of heating and to develop scientif-ically-grounded modes of technological treatment of titanium blanks [4].

Experiment

Material for research consisted of blades of steam turbines from two-phase titanium alloy VT6 (Ti-6Al-4 V) average durability after deformation in β—and final deformation by stamping in $\alpha+\beta$-regions near to resulting temperatures T_{50} and T_{70}, under the first and second technological circuits, accordingly. Contents researches of parameters of separate structural components are carried out on alloying ele-ments and a rating of their microhardness. Results have been received with use of methods of optical metallography, a transmission electronic microscopy and X-ray micro-spectral analysis. The temperature of final transition of an alloy in β-region (T_{pt}) made 1,015 °C.

A Structure and the Phase Composition of Blades Metal

In materials of steam blade turbines from alloy BV6, stamped under both techno-logical circuits, there was generated a structure of bimodal structures: α—the phase had both the form globular (α_I), and the form of extended plates (α_{II}), divided by interlayers β_{II}-phases, Fig. 12.2. Apparently from electron microscope pictures, Fig. 12.3, in a substructure of blades metal, received on the first technology, in comparison with the second, borders of all phase components α_I, α_{II}, β_{II} exhibited much better relaxation at the expense of dislocation of fine tunings, was less as was the number of the curved extinction contours, usually testifying to presence in the metal of internal stresses.

Fig. 12.2 The microstructure blades from alloy VT6 after stamping on the first (**a**) and on the second (**b**) technologies

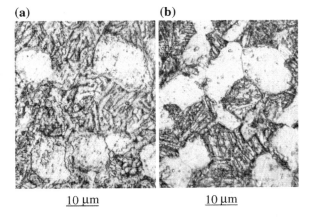

(a)   **(b)**

10 μm   10 μm

Fig. 12.3 The electron microscope structure of material blades from alloy VT6 after stamping on the first (**a**) and on the second (**b**) technologies

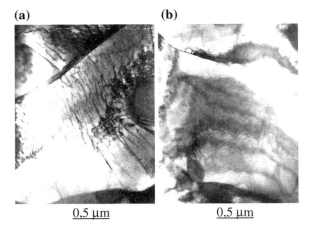

(a)   **(b)**

0,5 μm   0,5 μm

In Table 12.1 results of the sizes rating, the separate structural—phase components received with the help of an optical metallography and a transmission electronic microscopy in materials of steam blades turbines after stamping under two technological circuits are submitted.

As can be seen from Table 12.1, width of plates α_{II}-phases and thickness of β_{II}-layers after the first technology in comparison with the second, appeared more in 3 and 8 times, respectively. The ratio of width of plates α_{II}- of a phase and β_{II}-layers testifies to different temperature intervals of their formation and speeds of cooling. In blades received on the first technology at lower temperatures the structure of type Widmanstatten, which on the data from a number of authors, surpasses on fatigue strength a martensite-similar structure of blades metal received on the second technology [7], was generated. In materials of both blades with bimodal structure, the size lamellar $(\beta_{II} + \alpha_{II})$-a component was more size globular α_{I}-phases.

Table 12.1 The particles size of phase components of metal blades, fabricated on two technologies

Phase components	Size of particles of phases after the first technology (μm)	Size of particles of phases after the second technology (μm)
α_I	15	12
α_{II}	2.2	0.7
β_{II}	0.8	0.1
$\alpha_{II} + \beta_{II}$	22	18

Distribution of Alloying Elements in Metal of Blades

In materials of blades steam turbines from alloy VT6 fabricated stamping under two technological circuits, researches had been carried out on the contents of alloying elements (aluminum, vanadium, titanium, iron) in separate phase components, with the help of X-ray micro-spectral analysis (Table 12.2).

As results have shown, in both materials, plates secondary α_{II}-phases had a chemical composition, which approximately corresponded to an average composition of alloy Ti-6AL-4 V. In globalized primary α_I-phases, in comparison with plates α_{II}-phases, on the average contained a vanadium less on 2.2 % and aluminum more on 1.3 % (weight).

Narrow layers secondary β_{II}-phases have been enriched with a vanadium (β-stabilizing element).

And, in blades material fabricated on the first technology, in comparison with the second, β_{II}-the phase contained a vanadium more on 1.3 % and aluminum less on 0.5 % (weight).

Usually, such distribution of alloying elements provides relative "softness" of layers β_{II}-phase and higher operational properties of product material [8]. On the contrary, globalizes primary α_I-phases in blades material fabricated on the first technology, in comparison with the second, contained a vanadium less on 0.6 % and aluminum more on 1.3 % (weight).

On Fig. 12.4 typical diagrams of distribution of alloying elements with step 0.5 microns received with the help of X-ray micro-spectral analysis, in materials of turbine blades from alloy VT6 fabricated on both technologies are submitted. It can

Table 12.2 The contents of alloying elements in phase components of metal blades, fabricated on two technologies

Phase components	Concentration after the first technology (weight %)		Concentration after the second technology (weight %)	
	AL	V	AL	V
α_I	7.30	1.35	6.97	2.04
α_{II}	6.01	3.60	5.68	4.21
β_{II}	4.14	9.64	4.61	8.28

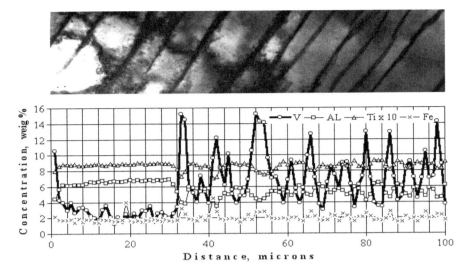

Fig. 12.4 The microstructure and distribution of alloying elements in structural components of material blades from alloy VT6

be seen that in a material, vanadium is the most non-uniformly distributed, being focused in β_{II}-phase. Its concentration changes from 1 up to 20 %. Aluminum is distributed more similarly, its concentration changes from 3 up to 7 % (weight).

A Rating of Microhardness of Metal of Blades

In materials of blades of steam turbines from alloy VT6 fabricated under two technological circuits, researches of microhardness of separate phase components have been carried out. In Table 12.3 the data of statistical treatment of results of measurements are submitted.

It can be seen that in blades material fabricated on the second technology, in comparison with the first, the level of microhardness of all phases was higher, which confirms presence of higher internal stresses on which it was already

Table 12.3 Microhardness of phase components of blades metal fabricated on two technologies

Phase components	Microhardness of particles of phases after the first technology (MPa)	Microhardness of particles of phases after the second technology (MPa)
α_I	3,550	3,664
α_{II}	3,830	4,091
β_{II}	2,584	3,000
$\alpha_{II} + \beta_{II}$	2,999	3,020

informed above (see "Introduction"). The blades material fabricated on both tech-nologies had a bimodal structure. Probably, the strength balance α_I- and ($\alpha_{II} + \beta_{II}$)-structural fashions can guarantee high serviceability of a material under loading. Apparently from Table 12.3, in blades materials fabricated on the first technology, in comparison with the second, the smaller difference in hardness between globular and lamellar structures, which made 551 and 644 MPa, accordingly has been achieved.

On Fig. 12.5 correlation dependences of results of measurements of microh-ardness on a chemical compound of separate phases (α_I, α_{II}, β_{II}) in a condition of delivery and in a martensite phase after hardenings from different temperatures are submitted. It can be seen that microhardness of plate's α_{II}-phases and, especially, globular α_I-particles grows with increase in the contents in them α—and, it is especial, β-stabilizing alloying elements. On the contrary, microhardness of β_{II}-layers considerably decreases with increase in the contents in the β-stabilizers. It is necessary to note, that microhardness of a martensite phase grows with increase in its contents of α-stabilizers, probably, forming α' —the martensite, and decreases with increase of β-stabilizing elements, forming α''—the martensite.

As it has been shown in "The Experimental Set", in blades material received on the first technology, in comparison with the second, the β-phase with big contents β—of stabilizing alloying elements was generated. It agrees with Fig. 12.5, such distribution of vanadium, really, provides rather smaller microhardness and big "softness" of layers β_{II}-phases.

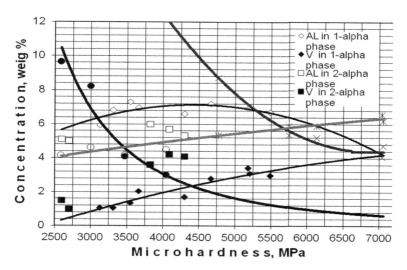

Fig. 12.5 Dependence of results of a rating of microhardness on a chemical compound of separate phase components (α_I, α_{II}, β_{II}, martensite) alloy VT6

Conclusion

As results of research have shown, blades materials of steam turbines from alloy VT6 fabricated by final stamping under two technological circuits had bimodal structure, in which the share lamellar ($\alpha_{II} + \beta_{II}$)-structures (50–70 %) prevailed of a share globular α_I-structures (30–50 %). The blades material fabricated on the first technology, in comparison with the second, possessed wider layers "soft" β_{II}-the phases, enriched same it β-stabilizers. At the same time, this material contained higher concentration of aluminum in primary globalizes α_I-phases, that provided strength balance (close microhardness) structural components.

It is known, that at cyclic loadings in regular intervals distributed soft faltering layers β_{II}-phases, transiting on the contour of rather solid secondary α_{II}-phases and of strength balance globalizes α_I-phases, complicating premature localization of plastic deformation and origin of a crack in separate phases. And at the stage of distribution, the crack is more difficult to increase the length in ($\alpha_{II} + \beta_{II}$)-lamellar structure as it is always braked by soft layers of β_{II}-phase and is compelled to change the trajectory, bending around globalized particles α_I-phases. Thus operational properties of product material increase [7–9].

A feature of effect of exposure steam drops loadings is not only recurrence and cyclicity of the enclosed stress, but also its dynamism. Though of short duration of influence, internal stresses have insufficient time to be redistributed, there is localization of the big loadings in small microvolumes, in separate structural and phase components of material. The successful combination of structure, chemical compound and properties, abilities to resist the microshock influence of these components, determines durability of products.

Researched alloy VT6 concerning martensite class, in stable condition contains 10–18 % β-phases, which at sharp cooling turns into α'- or α''-martensite. However, $\beta \rightarrow \alpha''$-transformation can take place and at room temperature. It is known that plastic deformation accelerates decomposition enriched β-stabilizers of β-solid solution with education α''-phases, as a result of a high level of internal stresses [10]. The subsequent ageing at temperatures 450–500 °C, results in its decomposition and education $\alpha' + (\beta)$-phases [11]. Presence of phase transformation $\alpha'' \rightarrow \alpha' + (\beta)$ results in significant strengthening of an alloy.

It is possible to believe, that in result of effect blows by pair, in soft enriched of vanadium, wide, regular intervals distributed β-layers there is an accumulation of defects of crystal structure and internal stresses. Transition deformed β-phases in a nonequilibrum condition, causes phase $\beta \rightarrow \alpha''$- transformation, accompanying with local frictional [12], and as consequence,—the subsequent ageing strengthening of microvolumes of alloy in result $\alpha'' \rightarrow \alpha' + (\beta)$—transformations. According to the generalized kinetic diagram displayed on Fig. 12.1, the more layers of β-phase contain quantity isomorphic β-stabilizing elements, those at lower temperatures and for a longer time will proceed with a $\beta \rightarrow \alpha''$-transformation (more completely).

Thus, it appears more significant that strengthening of borders between solid particles of α-phases and terms of operation of a finished article will increase, adding to the proof of the effect of exposure steam drops.

References

1. Deitch ME, Filippov GF (1987) Two-phase currents in elements of the heat power equipment. Monography Moscow Enegroatomizdat, 328 p (in Russian)
2. Skotnikova MA, Parshin AM (1997) Chart of disintegration and mode of a thermal obrakbotka of two-phase alloys of the titan. Metall Sci Heat Treat Met 1(7):31–37 (in Russian)
3. Skotnikova MA, Lanina AA, Krylov NA, Ivanov EK (2007) Structural and phase transformations in a blanket of a shovel of the steam turbine at a kapleudarny erosion. Sb. works. Problems of a resource and safe operation of materials SPb, pp 102–107 (in Russian)
4. Skotnikova MA, Ushkov SS (1999) Development of the scientific principle of a choice of final heat treatment of two-phase hot-rolled semi-finished products from alloys of the titan. Prog Mater Technol 3:91–98 (in Russian)
5. Skotnikova MA, Chizhik TA, Tsybulina IN, Lanina AA, Krylov NA (2007) Use of titanic alloys as a material of shovels of steam turbines. Mater Sci Quest 51(3):61–70 (in Russian)
6. Skotnikova MA, Lanina AA, Krylov NA, Homchenko EV (2010) Features of a structure of a material of shovels of steam turbines from a titanic alloy of VT6. Sb. works. Problem of materials science at design, production and operation of the equipment of the nuclear power plant. SPb Federal State Unitary Enterprise Central Research Institute KM Prometheus, pp 231–238 (in Russian)
7. Park J, Margolin H (1977) Metall Trans 15A(1):155–169
8. Holl I, Khammond K (1977) Proceedings conference Titanium-77, Moscow, Russia, vol 1, p 351
9. Skotnikova MA, Chizhik TA, Lisyansky AS, Simin ON, Tsybulina IN, Lanina AA (2009) Research of working shovels of turbines of big power taking into account structural and phase transformations in metal of stampings from a titanic alloy of VT6. Metalloobrabotka 54(6):12–21 (in Russian)
10. Gridnev VN, Ivasishin OM, Oshkaderov SP (1986) Physical bases high-speed thermo—hardenings of titanic alloys. Kiev, Ukraine: Naukova thought izdat, 256 p (in Russian)
11. Yermolov MI, Solonina OP (1967) Physics of metals and metallurgical science. 23(1):63–69 (in Russian)
12. Fedotov SG, Konstantinov KM (1987) New constructional material—the Titan. Monography. Moscow. Machine Industry Izdat, 220 p (in Russian)

Chapter 13
Conflicts in Product Development and Machining Time Estimation at Early Design Stages

Dmitry I. Troitsky

Abstract The paper considers the problem of efficiency and quality enhancement in computer-aided product development and manufacturing based on professional conflicts prevention between experts in different domains using a design model concept. The model enables manufacturability and logistical risk evaluation at the design stage and also ensures the decision's validity. A machining time evaluation model is also proposed.

Keywords Professional conflict · Machining time · Logistical risk · Product lifecycle management

Introduction

The modern CAD/CAE/CAM/PLM solutions are used at virtually every product lifecycle stage. However, there is a need for further investigation concerning the efficiency of collaboration between product and process development teams. The professional conflicts that inevitably arise in the course of product and process development lead to delays, and reduce the team efficiency.

A professional conflict (PC hereinafter) as considered in this paper is not something negative; it is an objective conflict not between two persons but between two actors [1, 2] in contrast to the conventional conflict definition used in conflict resolution studies. A product life cycle is a pipeline consisting of a number of stages. Each subsequent stage uses the information created at the previous stage. For example, product design is the initial data for production planning and raw materials purchasing. The subsequent actors validate the proposed design. Should the design fail to meet their requirements, a conflict occurs. To resolve such a conflict a number of iterations may be required. It causes extra and unnecessary costs, and efforts. In most cases a conflict is resolved through changing the design

D.I. Troitsky (✉)
Tula State University, Tula, Russia
e-mail: dtroitsky@tsu.tula.ru

© Springer International Publishing Switzerland 2015
A. Evgrafov (ed.), *Advances in Mechanical Engineering*, Lecture Notes in Mechanical Engineering, DOI 10.1007/978-3-319-15684-2_13

since the cost requirements (which define the manufacturability, materials intensity, and other key indicators) have already been set by the customer or aligned with the current market situation, and cannot be changed.

Computer-aided design tools have expanded the design documentation concept. Now it is a virtual model that includes both 3D geometry, and product and manufacturing information (PMI) like material properties, dimensional and geometric tolerances, surface finish requirements, etc.

The higher the PMI information content, the more the virtual model is suitable at the subsequent product lifecycle stages. For this reason a way of evaluating this information content is needed. As soon as PMI is available it would be possible to determine the derived design indicators (manufacturability, materials intensity, supply chain risks). These indicators provide a feedback to the designer to modify the design solution in such a way that the indicators are within the ranges specified by the subsequent product lifecycle actors.

When a complete virtual design, models, and algorithms for design indicators evaluation are available, any possible professional conflicts can be avoided since the reason for a conflict is eliminated. Such an approach would solve the most urgent manufacturing challenges such as reducing time to market, improving product development efficiency, and quality.

Problem Statement

The paper considers improving product development efficiency and quality through avoiding professional conflicts between the product and process development actors. It proposes a model forecasting the manufacturability and supply chain risk values at product design stage, and verifying the virtual design model's validity.

A product lifecycle is a sequence of procedures performed by different actors. Since the lifecycle is mostly a linear pipeline (attempts at concurrent product and process development to date have been mostly unsuccessful), the outcomes of an i-1th stage are the inputs for the ith stage. In this way the subsequent actors depend on the results obtained by the prior ones. Should an ith actor believe that the results obtained by one or several actors at $1...i-1$th stages are unsatisfactory, a professional conflict occurs. We propose the following definition:

> A professional conflict is an objective collision occurring as the actors (conflict entities) perform their professional duties. The reason for a conflict is different goals of each actor, and the interdependences of the actors as they make decisions.

Professional Conflicts Analysis

Let us consider a conflict between two actors A_{i-1} and A_i, where i is a sequential number of the actor's function in the product lifecycle. Generally an actor's function is choosing a specific design option from the SS phase space. SS is an

n-dimensional space; each dimension corresponds to one of the design variables. The space configuration is restricted. We have identified three kinds of such restrictions:

Restrictions of the 1st kind ($L1$): objective;
Restrictions of the 2nd kind ($L2$): functional
Restrictions of the 3rd kind (LA): lifecycle relationships.

Then there is the Sso subset of acceptable design options within the SS space:

$$SSo = SS - L1 - L2 - L3 \tag{13.1}$$

Obviously restrictions of the 1st kind do not lead to professional conflicts. Restrictions of the 2nd kind do not lead to conflict either since they are not results of other actors' activities but are defined with analysis or testing.

Restrictions of the 3rd kind are the most significant ones. When such restrictions are taken into account, the $i + 1$ actor has to act within the phase space of the $(1...i)$ actors, since these actors define the initial data for the $i + 1$ lifecycle stage. SSo (i-1) and $Sso(i)$ spaces must intersect. It should be noted that the $Sso(i)$ space configuration depends on the restrictions of the 3rd kind imposed by the specific design option P_{i-1} made at the $i-1$ stage. Making a design decision is choosing a specific design option from the phase space:

$$P_{i-1} = <p_1, p_2, \ldots p_n >, \tag{13.2}$$

where n is the number of dimensions in the SSo phase space;
p_i is the ith dimension.

The intersection condition is as follows:

$$Sso(i - 1) \cap Sso(i) \neq \emptyset. \tag{13.3}$$

If this condition is not met the corresponding conflict cannot be resolved at all since the restrictions applied to the configurations of the $SSo(i-1)$ and $Sso(i)$ spaces are contradictory.

If the condition is met, there are two possible situations: the $i-1$th design option either belongs or does not belong to the common area of the spaces. The former is expressed as:

$$P_{i-1} \in Sso(i - 1) \cap Sso(i). \tag{13.4}$$

The latter is:

$$P_{i-1} \notin Sso(i - 1) \cap Sso(i). \tag{13.5}$$

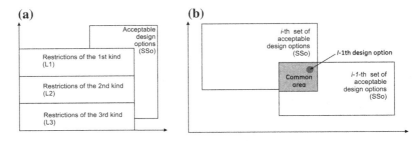

Fig. 13.1 Design restrictions (**a**), and the common area of two subsequent lifecycle stages (**b**)

When the condition (13.4) is met there is no conflict since the P_{i-1} design option is within the area acceptable for the ith actor. Otherwise an option acceptable for the $i-1$th actor happens to be unacceptable for the ith actor, and a professional conflict occurs (Fig. 13.1).

The perfect solution is meeting the condition (13.4) at every product lifecycle stage. To make the condition (13.4) true, the $i-1$th actor when making a decision has to consider the decision's consequences as perceived by the subsequent actors. It is clear that the actors lack the required information for such a consideration. In other words, the information entropy of design decisions is too high. So enhancing the information content of design decision and reducing its entropy is a way to not only resolve but also to avoid professional conflicts.

Any design option first appears as an intent to be represented with a formalized model including a geometry definition (a 3D model), and the product and manufacturing information (PMI) attributes: dimensional and geometric tolerances, material properties, surface finish, etc.)

The PMI attributes can be divided into two kinds.

Attributes of the 1st kind are specified by the designer considering the product requirements, the designer's experience, regulatory requirements, etc. Today most CAD systems support linking such attributes to 3D geometry. The availability of such attributes increases the information content of a design decision, and decreases its entropy. In order to avoid conflicts we introduce the "design model" concept. A design model also includes attributes of the 2nd kind: the information required to evaluate and validate the design at subsequent lifecycle stages. The most significant attributes of the 2nd kind are manufacturability, supply chain risks (these are two most important contributors to product cost), and the validity of the design model itself.

We propose the following definition (Fig. 13.2):

Design model is a set of information that includes the geometry of the object being designed, its attributes of the 1st kind linked to the geometry, and its attributes of the 2nd kind and some extra information as needed to evaluate these attributes (regulatory requirements, references, statistical data for a specific manufacturing facility, etc.)

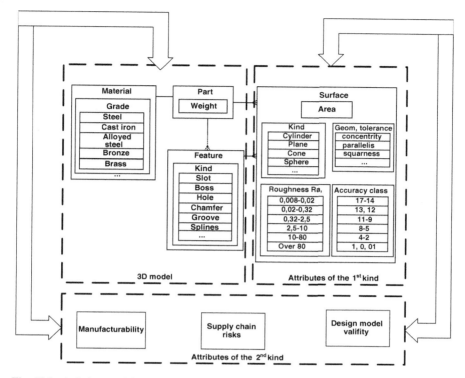

Fig. 13.2 A design model

This paper considers the following attributes of the 2nd kind:

A21:   machining time, minutes;
A22:   an integrated supply chain risk indicator ([0–1] range);
A23:   the design model validity indicator ([0–1] range).

Thus a design model can be represented as the following tuple:

$$DM = <G, R, A1, A2> \quad\quad\quad\quad (13.6)$$

where G is a set of geometric entities;
R is a set of relations between the geometric entities;
$A1$ is a set of attributes of the 1st kind;
$A2$ is a set of attributes of the 2nd kind.

Figure 13.3 shows an IDEF diagram of the proposed model's application to the product development processes.

Part Manufacturability Analysis at Design Stage. Due to the limited size of the paper here we consider only one of the three attributes of the 2nd kind: machining time [7]. Machining time is the primary contributor to product cost [3].

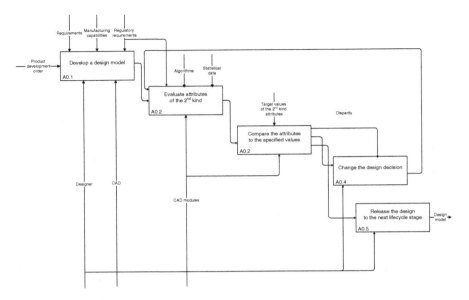

Fig. 13.3 An IDEF diagram of the proposed model

Machining time evaluation at early design stages is the best way to avoid professional conflicts.

The only feasible method for machining time evaluation is regression analysis [4]. It means the development of a polynomial equation that expresses the relation between attributes of the 1st kind (available in a design model) and machining time [8]. The machining time evaluation workflow is shown in Fig. 13.4.

The proposed workflow includes further model refinement through a feedback to the manufacturing. Actual machining time values are fed back into the self-learning module which recalculates the regression equation factors.

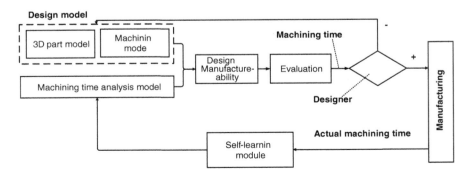

Fig. 13.4 Machining time evaluation workflow

The proposed multiple regression equation is as follows:

$$T = F\left(M, \sum_{i=1}^{n} f(S_i, Ra_i, t_i)\right) \cdot k_{MAT} \tag{13.7}$$

where M is the part weight; S_i is the surface of the ith surface being machined; Ra_i is the specified roughness of the ith surface; t_i is the dimensional tolerance of the ith surface, k_{MAT} is the material machinability factor.

Let K be as follows:

$$K = \sum_{i=1}^{n} f(S_i, Ra_i, t_i). \tag{13.8}$$

Then the machining time evaluation model is:

$$\begin{cases} T = F(M, K) \\ K = \sum_{i=1}^{n} f(S_i, Ra_i, t_i) \end{cases} \tag{13.9}$$

The problem is to find the relation between K and S, Ra, t from a matrix of experimental data:

$$\begin{vmatrix} S_1 & Ra_1 & t_1 & K_1 \\ S_2 & Ra_2 & t_2 & K_2 \\ \vdots & \vdots & \vdots & \vdots \\ S & Ra & t & K \end{vmatrix}. \tag{13.10}$$

The sought expression for K is a power-law dependence [5]:

$$K = a_0 \cdot S^{x_1} \cdot Ra^{x_2} \cdot t^{x_3}, \tag{13.11}$$

where a_0 is a correlation factor.

Equation (13.11) can be converted to a linear form:

$$\lg K = \lg a_0 + x_1 \cdot \lg S + x_2 \cdot \lg Ra + x_3 \cdot \lg t. \tag{13.12}$$

The following notation shall be used:

$$\begin{aligned} &\lg K = Y, \lg a_0 = a_0', \lg S = U_1, \lg Ra = U_2, \lg t = U_3, \\ &Y = a_0' + x_1 \cdot U_1 + x_2 \cdot U_2 + x_3 \cdot U_3. \end{aligned} \tag{13.13}$$

By taking a logarithm of each element of the matrix (13.10) we obtain

$$
\begin{vmatrix}
\lg S_1 & \lg Ra_1 & \lg t_1 & \lg K_1 \\
\lg S_2 & \lg Ra_2 & \lg t_2 & \lg K_2 \\
\vdots & \vdots & \vdots & \vdots \\
U_1 & U_2 & U_3 & Y
\end{vmatrix}
\tag{13.14}
$$

The entire system of equations is:
The unknown values are calculated as follows:

$$
\sum U_1 = \lg S_1 + \lg S_2 + \ldots + \lg S_n,
$$
$$
\sum U_2 = \lg Ra_1 + \lg Ra_2 + \ldots + \lg Ra_n
$$
$$
\sum U_3 = \lg t_1 + \lg t_2 + \ldots + \lg t_n,
$$
$$
\sum U_1^2 = (\lg S_1)^2 + (\lg S_2)^2 + \ldots + (\lg S_n)^2,
$$
$$
\sum U_2^2 = (\lg Ra_1)^2 + (\lg Ra_2)^2 + \ldots + (\lg Ra_n)^2,
$$
$$
\sum U_3^2 = (\lg t_1)^2 + (\lg t_2)^2 + \ldots + (\lg t_n)^2,
$$
$$
\sum U_1 U_2 = \lg S_1 \cdot \lg Ra_1 + \lg S_2 \cdot \lg Ra_2 + \ldots \lg S_n \cdot \lg Ra_n,
$$
$$
\sum U_1 U_3 = \lg S_1 \cdot \lg t_1 + \lg S_2 \cdot \lg t_2 + \ldots \lg S_n \cdot \lg t_n,
$$
$$
\sum U_2 U_3 = \lg Ra_1 \cdot \lg t_1 + \lg Ra_2 \cdot \lg t_2 + \ldots \lg Ra_n \cdot \lg t_n.
$$

Solving the system with the method of successive elimination the following expressions are obtained:

$$
a_0' = \frac{1}{4} \left[1 - \frac{0.5 \cdot (\sum U_1)^2 - 1/8a \cdot \sum U_1 \cdot \sum U_2}{0.25 \cdot (\sum U_1)^2 - \sum U_1^2} \right.
$$
$$
- \frac{1.25 (\sum U_2)^2}{a (\sum U_1 \sum U_2 - 4 \sum U_1 U_2)}
$$
$$
\left. - \frac{\sum U_2 \sum U_3 + a \sum U_1 \sum U_3 + 4a \sum U_1 \sum U_3}{4a \sum U_1 U_3 - a \sum U_1 \sum U_3} \right];
\tag{13.15}
$$

$$
x_1 = \frac{0.5 \cdot \sum U_1 - 0.125a \cdot \sum U_2}{0.25 \cdot (\sum U_1)^2 - \sum U_1^2};
\tag{13.16}
$$

$$
x_2 = \frac{0.25 \sum U_2 + \sum U_2}{a \cdot (\sum U_1 \sum U_2 - 4 \sum U_1 U_2)};
\tag{13.17}
$$

$$
x_3 = \frac{\sum U_2 + a \cdot (\sum U_1 + 4 \sum U_1)}{a \cdot (4 \sum U_1 U_3 - \sum U_1 \sum U_3.)}
\tag{13.18}
$$

In (13.15–13.18) the a value is:

$$a = \frac{0.25 \sum U_1 \sum U_2 - \sum U_1 U_2}{\left(\sum U_1\right)^2 - 4 \sum U_1^2}. \qquad (13.19)$$

Now we determine the relation between T (machining time), K and M:

$$T = f(M, K).$$

Similarly we obtain a matrix:

$$\begin{vmatrix} M_1 & K_1 & T_1 \\ M_2 & K_2 & T_2 \\ : & : & : \\ M_n & K_n & T_n \end{vmatrix}; \qquad (13.20)$$

$$T = b_0 \cdot M^{y_1} \cdot K^{y_2}; \qquad (13.21)$$

$$\lg T = \lg b_0 + y_1 \cdot \lg M + y_2 \cdot \lg K; \qquad (13.22)$$

$$Z = b_0' + y_1 \cdot P_1 + y_2 \cdot P_2; \qquad (13.23)$$

$$\begin{vmatrix} \lg V_1 & \lg K_1 & \lg T_1 \\ \lg V_2 & \lg K_2 & \lg T_2 \\ : & : & : \\ P_1 & P_2 & Z \end{vmatrix};$$

$$\begin{cases} 1 = 3b_0' + y_1 \cdot \sum P_1 + y_2 \cdot \sum P_2 \\ \sum P_1 = b_0' \cdot \sum P_1 + y_1 \cdot \sum P_1^2 + y_2 \cdot \sum P_1 P_2 \\ \sum P_2 = b_0' \cdot \sum P_2 + y_2 \cdot \sum P_2^2 + y_1 \cdot \sum P_1 P_2 \end{cases} \qquad (13.24)$$

$$\sum P_1 = \lg V_1 + \lg M_2 + \ldots + \lg M_n;$$

$$\sum P_2 = \lg K_1 + \lg K_2 + \ldots + \lg K_n;$$

$$\sum P_1^2 = \lg M_1^2 + \lg M_2^2 + \ldots + \lg M_n^2;$$

$$\sum P_2^2 = \lg K_1^2 + \lg K_2^2 + \ldots + \lg K_n^2;$$

$$\sum P_1 P_2 = \lg M_1 \cdot \lg K_1 + \lg M_2 \cdot \lg K_2 + \ldots + \lg M_n \cdot \lg K_n.$$

The solution of (13.24) is:

$$y_2 = \frac{\frac{2}{3}\sum P_2 - c \cdot \sum P_1 + \frac{c}{3} \cdot \sum P_1}{\frac{c}{3} \cdot \sum P_1 \sum P_2 - c \cdot \sum P_1 P_2 + \sum P_2^2 - \frac{1}{3} \cdot \left(\sum P_2\right)^2}; \qquad (13.25)$$

$$y_1 \frac{\frac{2}{3} \cdot \sum P_1 - y_2 \cdot \left(\sum P_1 P_2 - \frac{1}{3}\sum P_1 \sum P_2\right)}{\sum P_1^2 - \frac{1}{3}\left(\sum P_1\right)^2}; \qquad (13.26)$$

$$b_0' = \frac{1}{3} \cdot \left(1 - y_1 \cdot \sum P_1 - y_2 \cdot \sum P_2\right), \qquad (13.27)$$

$$c = \frac{0.25 \sum P_1 \sum P_2 - \sum P_1 P_2}{\left(\sum P_1\right)^2 - 4\sum P_1^2}. \qquad (13.28)$$

Virtual Experiments

To obtain actual equations for machining time estimation we need a large amount of valid manufacturing data. Unfortunately, in real-life production the machining time is not always estimated correctly. We have analyzed the data available at three manufacturing companies and have found numerous errors. For this reason it has been decided to perform a number of virtual machining runs in a CAM system. We have studied milling using Delcam PowerMill and SPRUT CAM. Both systems indicate actual machining time as a by-product when developing and verifying NC codes.

We have deducted the following generalized regression equations that express the relations between machining time T, surface area F, and cutting feed Sp [6]:

$$T(Sp, F) = 3x^3 + 3y^3 + 3z^3 + 4.304x - 0.061y;$$
$$T(Sp, t) = 3x^3 + 3y^3 + 3z^3 - 0.121y^2 + 1.497z^2 - 6.422xy + 0.016yz + 12.806x - 0.062;$$

$$T(F, t) = 3x^3 + 3y^3 + 3z^3 + 0.979x - 4.695y + 3.452z - 0.519.$$

The graphs for these equations are shown in Fig. 13.5.

The following regression equations have been produced for primary milling features (Table 1):

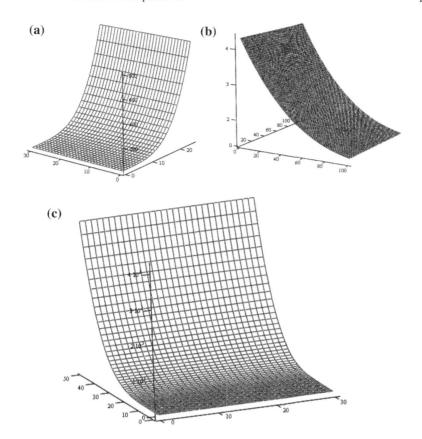

Fig. 13.5 Generalized machining time graphs: **a** $T(F,t)$, **b** $T(S,F)$, **c** $T(S,t)$

Table 1 Machining Time Equations for Different Part Features	No.	Feature	Equation
	1	Flat area	$T = 10^{4.07} \cdot S^{-0.6} \cdot Ra^{-0.25} \cdot t^{-1.07}$
	2	Cylindrical surface	$T = 10^{4.07} \cdot S^{-0.6} \cdot Ra^{-0.25} \cdot t^{-1.07}$
	3	Key slot	$T = 10^{-3.04} \cdot S^{0.098} \cdot t^{-0.34} \cdot d^{3.67}$
	4	Blind hole	$T = 10^{4} \cdot S^{-0.84} \cdot Ra^{-0.11} \cdot t^{-0.92}$
	5	Through hole	$T = 10^{4.03} \cdot S^{-0.68} \cdot Ra^{-0.34} \cdot t^{-0.72}$
	6	Slot	$T = 10^{4.01} \cdot S^{-0.65} \cdot Ra^{-0.14} \cdot t^{-1.27}$
	7	Boss	$T = 10^{4} \cdot S^{-0.77} \cdot Ra^{-0.31} \cdot t^{-0.14}$

Machining Time Model Verification

To validate the proposed model we will apply it to estimate the machining time for milling the part shown in Figs. 13.6 and 13.7.

The part has 16 surfaces to be machined (see Fig. 13.7): two holes, three slotted surfaces, and nine planes. They are listed in the Table 2:

The total machining time is a sum of machining times for the holes, the slot, and the other surfaces. Surfaces 1 and 10, 2 and 11 are jointly machined.

Then

$$T_{1,10} = T_{2,11} = 10^{2.07} \cdot 150^{-0.6} \cdot 0.63^{-0.25} \cdot 5^{-1.07} = 1.1658 \, \text{min};$$

$$T_{15} = T_{16} = 10^{2.01} \cdot 100^{-0.65} \cdot 1,25^{-0.14} \cdot 5^{-1.27} = 0.6438 \, \text{min};$$

$$T_{12-14} = 10^{2.03} \cdot 150^{-0.68} \cdot 1,25^{-0.34} \cdot 6^{-0.72} = 0.9058 \, \text{min};$$

$$T_4 = 10^{2.07} \cdot 100^{-0.6} \cdot 0.63^{-0.25} \cdot 2^{-1.07} = 3.9634 \, \text{min};$$

$$Y = T_{1,10} + T_{2,11} + T_{15} + T_{16} + T_4 = 8.4884 \, \text{min}.$$

$$T_{PART} = 10^{-1.034} \cdot M^{-0.202} \cdot Y^{1.76} \cdot k_{MAT};$$

$$M = 0.144 \, \text{kg}, \ k_{MAR} = 0.8$$

$$T_{PART} = 4.718 \, \text{min}.$$

Fig. 13.6 A part manufactured by Milling (courtesy OAO TulaTochMash, Tula, Russia)

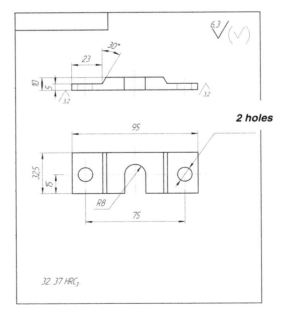

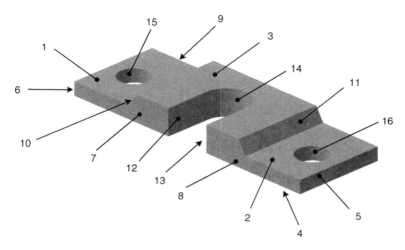

Fig. 13.7 3D model of a part manufactured by milling (courtesy OAO TulaTochMash, Tula, Russia)

Table 2 Part Surface Parameters

No.	Surface type		Surface area, mm²		Roughness
1	Plane		652.47		Ra 3.2
2			652.47		
3			1044.33		Ra 6.3
4			2536.902		Ra 3.2
5			162.5		Ra 6.3
6			162.5		
7			272.78		
8			272.78		
9			705.57		
10			187.64		
11			187.64		
12	Slot	Plane	162.5	576.33	
13		Plane	162.5		
14		Cylinder	251.33		
15	Cylinder (hole)		172.79		
16	Cylinder (hole)		172.79		

The machining time estimated with the available reference tables is 5.2 min. The difference is about 11 %, so the proposed model is valid and feasible.

Conclusion

The proposed machining time estimation model can be used to avoid professional conflicts in product development. Other attributes of the 2nd kind are evaluated with another two models that go beyond the scope of this paper. Further research will include creating models for other manufacturing methods (turning, stamping) and developing a CAD module/library to estimate the machining part from a 3D model and its PMI attributes.

References

1. Saaty T (1968) Mathematical models of arms control and disarmament.Wiley, New York
2. Saaty T, Vargas LG (1994) Decision making in economic, social and technological environments. RWS, Pittsburgh
3. Kaczmarek J (1976) Principles of machining by cutting, abrasion and erosion. Peter Peregrinus Ltd., Stevenage
4. Hazewinkel Michiel (ed) (2001) Regression analysis, encyclopedia of mathematics. Springer, Berlin
5. John E. Neely, Richard R. Kibbe, Roland O. Meyer, Warren T. White (2005) Machine Tool Practices. Prentice Hall PTR, 864 p
6. Inozemtsev AN, Novikova MV, Troitsky DI (2007) Manufacturability analysis in intelligent CAD Systems. In: 10th international conference computer graphics and artificial intelligence. Proceedings of the conference. Athens, Greece, pp 295–298
7. Bajaj M (2003) Towards next-generation design-for-manufacturability (DFM) frameworks for electronics product realization In: M. Bajaj, R. Peak, M. Wilson (eds) International Electronics Manufacturing Technology Symposium
8. Liua C, Lia Y, Wanga W, Shenb W (2013) A feature-based method for NC machining time estimation. Robot Comput-Integr Manuf 29(4):8–14

Chapter 14
Vibrations Excitation in Cyclic Mechanisms Due to Energy Generated in Nonstationary Constraints

Iosif I. Vulfson

Abstract The exchange of vibrational energy between an external source and subsystems of the cyclic mechanism schematized as a dynamic model with slowly varying parameters is studied. The conditions are obtained whose violation yields the formation of regions where the energy of vibrations appears in the time interval of the kinematic cycle, even in the absence of external perturbations. It was shown that this effect occurs due to the operation of the external source upon implementation of nonstationarity of dynamic constraints and the dynamic instability of the system on a finite time interval. The dynamic effects and key conclusions are illustrated by results of computer simulation. Engineering recommendations for decreasing the vibrational activity of cyclic mechanisms are presented.

1. In solving problems of machine dynamics we are often faced with a variety of effects that are generated due to energy transfer from an external source, as well as from one subsystem to another or energy exchange between different oscillations shape modes. Sometimes these effects are positive and facilitate vibrational protection of machines and mechanisms. One of the most widely known and vivid examples of this effect is dynamic damping. Under proper tuning the reaction from the side of the damper onto the main mass in the steady-state mode is equal to the disturbing force, but with opposite sign. In this case, the energy of the external source is purposely transferred from the object under vibrational protection to the dynamic damper. The analogous effect is observed upon dynamic unloading of the drive of cyclic mechanisms where the energy exchange between the effector and the dynamic unloader takes place [1, 2].

In other cases, these effects yield an undesirable redistribution of vibrations and their localization uncertain units, links, cross sections, etc. Here, a large role is often played the nonstationarity of dynamic constraints due to the variability of the vibratory system's parameters, that is inherent for the drives of machines with cyclic mechanisms. In particular, it is quite possible that local disturbances of the

I.I. Vulfson (✉)
Saint Petersburg State University of Technology and Design, Bolshaya Morskaya str. 18, 191186 Saint Petersburg, Russia
e-mail: jvulf@yandex.ru

© Springer International Publishing Switzerland 2015
A. Evgrafov (ed.), *Advances in Mechanical Engineering*, Lecture Notes in Mechanical Engineering, DOI 10.1007/978-3-319-15684-2_14

dynamic stability conditions, when the energy "replenishment" of the system yields an intense increase in the vibration amplitudes, take place at limited time intervals [1, 3–5]. Spatial localization in vibratory chains [6–8], when strict dynamic regularity of the system is violated in certain sections due to so-called "inclusions" that manifest themselves as pronounced extrema in the forms of vibration, belongs to this class of problems. Based on the classical model, this problem is defined concretely in [6] by the example of vibrations of a string and Bernoulli–Euler and Timoshenko beams with concentrated inclusions.

Similar effects are also observed upon the analysis of the dynamics of machines and automated lines with repetitive sections (modules). In particular, elimination of spatial localization is necessary in designing machines with an increased extension of the region of technological treatment of items when vibrations of long actuators should be similar to in-phase ones [1, 3–5]. The violation of this requirement yields (apart from undesirable dynamic effects) different flaws in the products, e.g., nonuniformity of the yarn and selvage defects in the manufacture of fabrics, thread breaks, defects of printed output in printing machines, violation of the specified accuracy and purity of treated surfaces in metal-cutting machines, etc. Preservation of the inphase vibration forms over a relatively large number of cyclic drive mechanisms is a rather complicated problem that requires further investigation.

The analysis showed that, as applied to the problems of machine dynamics, this problem is beyond the framework of classical theory. This is related to the fact that repetitive modules which form complicated systems with variable parameters and nonlinear elements appear instead of the point masses. In addition, the "inclusions", which are attributed to the vibratory systems of machines with deviations from regularity due to design and other factors that break the strict dynamic identity of the repetitive modules, are drastically different. Note that the problem analyzed here is quite general and it can be encountered in many physical problems. For example, the so-called Landau–Zener tunneling, in which the energy exchange takes place between two levels, is known in quantum mechanics [9, 10]. A mechanical analogue of tunneling, which represents a system of two weakly coupled pendulums where the partial frequency of one slowly varies in time and covers the region of internal resonance, was proposed in [10]. In this case, if the transfer of energy from one vibrator to another one turns out to be irreversible, there appears a trap for "capturing" the vibratory frequency. This effect, illustrated in Fig. 14.1 by the plots

Fig. 14.1 Energy transfer in coupled oscillators with varying parameters

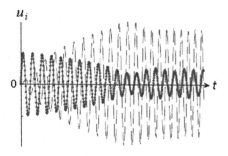

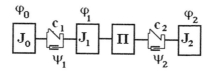

Fig. 14.2 Dynamic model

showing the vibration transfer from one subsystem (solid line) to another (dashed line), is presented in [9].

Let us consider the transfer of vibrational energy by the example of the dynamic model of a cyclic mechanism with an elastic drive (Fig. 14.2).

We use the following notation: J_0 is the moment of inertia; c_i are the coefficients of rigidity; ψ_i are the scattering coefficients; $\varphi_1 = \varphi_0 + q_1$; $\varphi_2 = \Pi(\varphi_0) + q_2$, where φ_i are the absolute angular coordinates of the corresponding inertial elements; $\varphi_0 = \omega_0 t$ is the ideal coordinate of the element J_0 at $\omega_0 = const$; q_i are the generalized coordinates which are equal to the absolute dynamic errors, i.e., deviation from the specified programmed motion; Q_i are the generalized forces.

Assuming the function $\Pi(\varphi_1)$ to be continuous and differentiable, let us linearize this function and its first two derivatives in the vicinity of the programmed motion [1, 2]:

$$\Pi(\varphi_0 + q_1) \approx \Pi_* + \Pi'_* q_1; \quad \Pi'(\varphi_0 + q_1) \approx \Pi'_* + \Pi''_* q_1, \tag{14.1}$$

where the asterisk corresponds to the argument $\varphi_0 = \omega_0 t$; $(\ )' = d/d\varphi$.

According to Eq. (14.1) the following set of differential equations corresponds to this model after linearization in the vicinity of the programmed motion:

$$\ddot{q}_1 + k_1(2\delta_1 \dot{q}_1 + k_1 q_1) + \mu k_2 \Pi'_* [2\delta_2(\Pi'_* \dot{q}_1 - \dot{q}_2) + k_2(\mu \Pi'_* q_1 - q_2)] = W_1(t);$$
$$\ddot{q}_2 + k_2[2\delta_2(\dot{q}_2 - \Pi'_* \dot{q}_1) + k_2(q_2 - \Pi'_* q_1)] = W_2(t), \tag{14.2}$$

where $k_i = \sqrt{c_i/J_i}$, $\mu = \sqrt{J_2/J_1}$, $\delta_i = \psi_i/(4\pi)$ at $(i = 1, 2)$; and $W_i(t)$ is the external excitation.

To understand the observed dynamic effects, let us consider the particular case when the problem is reduced to the analysis of the vibratory system with variable parameters and one degree of freedom. If we schematize the output unit of the cyclic mechanism as a perfectly rigid body ($c_2 \rightarrow \infty$), the system is described with the following nonlinear differential equation:

$$J_1 \ddot{q}_1 + c_1 q_1 = -\Pi'(\varphi_1)\{J_2[\Pi''(\varphi_1)(\dot{q}_1 + \omega_0)^2 + \Pi'(\varphi_1)\ddot{q}_1] + Q_1^*\}, \tag{14.3}$$

where, in addition to the introduced denomination, we assume that $q_1 = \varphi_1 - \varphi_0$, Q^* is the external nonconservative generalized force.

According to Eqs. (14.1) and (14.3) the linearized differential equation in this case takes on the form

$$\ddot{q}_1 + 2n(t)\dot{q}_1 + p^2(t)q_1 = W(t), \tag{14.4}$$

where $p = k_1/\sqrt{1 + \mu^2\Pi_*'^2}$, $k_1 = \sqrt{c_1/J_1}$; the function $n(t)$ consists of the dissipative and gyroscopic components $n(t) = n_0(t) + n_1(t)$.

It can be shown that $n_0(t) = \delta_1 p(t)$, where $\delta = \psi_1/(4\pi)$, ψ_1 is the coefficient of scattering at given the positional dissipative force. Distinguishing in Eq. (14.3) the terms which are proportional to vibration velocity we have

$$n_1 = \omega_0\mu^2\Pi_*'\Pi_*''/(1 + \mu^2\Pi_*'^2) \tag{14.5}$$

Note that Eq. (14.4) is valid for any vibratory system with one degree of freedom which describes a drive with a reduced moment of inertia $J(\varphi) = J_1(1 + \mu^2\Pi'^2)$, where $\mu^2 = (J_{max} - J_1)/J_1$ or with the variable reduced coefficient of rigidity [2].

Let us make use of the method of conditional vibrator which is a slowly varying function of position that yields the solution at the level of the WKB approximation of the first order [1, 2]. Then

$$q_1 = A_0\exp[-\int_0^t n(\xi)d\xi]\sqrt{k_1/p(t)}\cos[\int_0^t p(\xi)d\xi + \alpha] + Y(t), \tag{14.6}$$

where A_0, α are determined by the initial conditions, and the partial solution $Y(t)$ is defined as

$$Y(t) = \frac{1}{\sqrt{p(t)}}\int_0^t \frac{W(u)}{\sqrt{p(u)}}\exp[-\int_u^t n(\xi)d\xi][\sin\int_u^t p(\xi)d\xi]du. \tag{14.7}$$

At $p = \text{const}$, dependence (14.7) coincides with the Duhamel formula.

In the vibratory system analyzed, the energy exchange with the external energy source of unlimited power occurs apart from the energy losses from overcoming the dissipative forces.

Let us restrict ourselves to the analysis of free vibrations ($Y(t) \equiv 0$). In the practice of dynamic calculation of machines, free vibrations are often of interest from the standpoint of frequency analysis which is an important stage in the estimation of forced vibrations. In this case, it is taken into account that free vibrations formed due to the energy introduced into the system at the initial instant are damped fairly quickly and, consequently, do not actually affect the steady state vibratory modes. Meanwhile, the vibrational activity of cyclic machines mainly depends on the level of the so-called free accompanying vibrations [1–5, 11] which, with complex laws of motion, clearances, and other perturbations of a pulsed nature, do not only damp, but also excite. The key sources of excitation of such vibrations in cyclic mechanisms are discontinuous or sharp variations in the derivatives of the function of position. Strictly speaking, these vibrations should be referred to the class of forced vibrations; however, from the standpoint of the frequency spectrum, remoteness from resonances, and methodical considerations it is more convenient to consider them as free vibrations appearing at $t = t_i > 0$. In this case, the corresponding "initial" conditions are determined when partial solution (14.7) is represented as a rapidly convergent series over the derivatives of the perturbation function $W(t)$ [1, 2].

Let us single out the variable component of the vibrational amplitude which, according to Eq. (14.6), is described by the function

$$S = \exp[-\int_0^t (n_0 + n_1)dt] \sqrt{k_1/p(t)}. \tag{14.8}$$

According to Eq. (14.7), at $dS/dt < 0$ the free vibrations of the system at any t will be decreasing. In this case, according to Eq. (14.8) the following condition should be met [1, 2] :

$$n + \dot{p}/(2p) > 0. \tag{14.9}$$

It is interesting that condition (14.9) can be obtained at arbitrary variations in $p(t)$ based on the direct Lyapunov method that sets sufficient conditions of asymptotic stability. Actually, if we take the square of the amplitude of free vibrations as the Lyapunov function, we have $U = q^2 + (\dot{q}_1/p)^2$. Omitting the computations, we can write as follows

$$\dot{U} = \dot{q}_1^2 [\frac{d}{dt}(p^{-2}) - 4n/p^2]. \tag{14.10}$$

According to the second Lyapunov theorem, the sufficient condition for the asymptotic stability has the form $\dot{U} < 0$. One can easily ensure that when Eq. (14.10) is taken into account this condition coincides with condition (14.9). This means that for estimating the amplitudes at slow variation in the parameters at the level of the WKB approximation of the first order, condition (14.9) is not only sufficient, but also necessary. Based on Eq. (14.10), for the analyzed model we obtain

$$\delta_1 > \delta_1^* = \frac{\mu^2 \Pi_*' \Pi_*'' \omega_0}{2k_1 \sqrt{1 + \mu^2 \Pi_*'^2}}. \tag{14.11}$$

Bearing in mind that the parameter δ_1 is proportional to the scattering coefficient ψ_1, this condition points to the fact that the variability of the system parameters does not only change the intensity of damping of free vibrations, but can also lead to "negative" damping, i.e., to substantial qualitative changes in the vibratory process. It follows from Eq. (14.11) that the amplitude increase should be expected at the braking sections when the kinetic power proportional to the product $\Pi'_*\Pi''_*$ is negative. At the constant value of the first transfer function $\Pi'_* = \text{const}$, we have $\Pi'' = 0$ and $\delta_1^* > 0$ that, as expected, corresponds to damped free vibrations during the entire kinematic cycle.

To estimate the energy variation intensity let us make use of the function

$$E = p^2(t)S^2(t), \tag{14.12}$$

to which the amplitude values of the energy are proportional.

Based on Eqs. (14.8) and (14.12), the condition $dE/dt < 0$ yields the following result:

$$n - \dot{p}/(2p) > 0. \tag{14.13}$$

According to Eqs. (14.5) and (14.13), the condition which is analogous to relation (14.11) takes on the form

$$\delta > \delta_* = \frac{3\mu^2 \Pi'_*\Pi''_*\omega_0}{2k_1\sqrt{1 + \mu^2\Pi'^2_*}}. \tag{14.14}$$

Figure 14.3 shows the family of critical values of the $\delta_*(\varphi_0, \mu^2)$ parameter. The energy of vibratory systems increases in the process of acceleration ($\Pi'_*\Pi''_* > 0$) and decreases during deceleration ($\Pi'_*\Pi''_* < 0$). As μ approaches unity, the intensity of vibrational energy variations decreases.

Let us now return to the analysis of the initial model with two degrees of freedom (Eq. 14.2). As a rule, the parameter n in problems of the dynamics of mechanisms negligibly affects the "natural" frequency p and, at the same time,

Fig. 14.3 Critical values of the parameter of dissipation

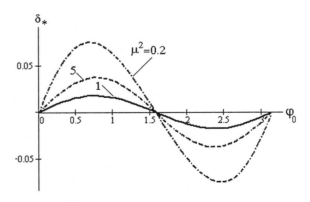

substantially influences the vibration frequencies. The analogous situation is identified when analyzing the systems with variable parameters which are reflected in models with many degrees of freedom. Then, with discretely specified parameters after linearization in the vicinity of the programmed motion, the frequency and modal analysis is based on the set of homogeneous differential equations

$$\mathbf{a}(t)\,\ddot{\mathbf{q}} + \mathbf{c}(t)\mathbf{q} = 0 \tag{14.15}$$

where $\mathbf{a}(t)$, $\mathbf{c}(t)$ are the square matrices of inertial and quasielastic coefficients, and q is the vector function of the generalized coordinates.

It was shown in [12] that the kinetic and potential energies are determined (accurate within the magnitude of the first order) by dependences

$$T = 0.5 \sum_{r=1}^{H} a_r^* \dot{\eta}_r^2; \quad V = 0.5 \sum_{r=1}^{H} c_r^* \eta_r^2,$$

where η_r are quasinormal coordinates; H is the number of degrees of freedom of the oscillatory system.

The variable "natural" frequencies in the first approximation can be defined on the basis of formal frequency equation, in which time plays a parameter role:

$$\det\left(c_{ij}(t) - a_{ij}p(t)^2\right) = 0. \tag{14.16}$$

In quasinormal coordinates, the set of equations takes on the form

$$a_r^*(t)\,\ddot{\eta}_r + [b_r(t) + \dot{a}_r^*(t)]\,\dot{\eta}_r + c_r^*(t)\eta_r = M_r(t), (r = \overline{1, H}) \tag{14.17}$$

where $M_r(t) = \sum_{i=1}^{H} \alpha_{ir} Q_{ir}^*$; Q_{ir}^*, is the nonconservative generalized force from external loads and kinematic excitation; $b_r(t)$ is the coefficient of the equivalent linear resistance reduced to the form r;

$$a_r^* = \sum_{i=1}^{H} \sum_{j=1}^{H} \alpha_{ir}\alpha_{jr}a_{ij}; \quad c_r^* = \sum_{i=1}^{H} \sum_{j=1}^{H} \alpha_{ir}\alpha_{jr}c_{ij},$$

where α_{ij} are the nonstationary shape factor.

In matrix form we have

$$\boldsymbol{\alpha}^T \mathbf{a}\, \boldsymbol{\alpha} = \mathrm{diag}\{a_1^*, \ldots a_H^*\}; \quad \boldsymbol{\alpha}^T \mathbf{c}\, \boldsymbol{\alpha} = \mathrm{diag}\{c_1^*, \ldots c_H^*\}$$

In the given method for determining quasinormal coordinates, the forms of vibrations (based upon physical prerequisites) are assumed to be slowly varying functions. As for the rest, their determination does not differ from the analogous procedure with constant parameters. Note that, as distinct from the traditional frequency indexing when a higher frequency corresponds to a larger index, the indexing at which $k_r = \lim\limits_{\Pi' \to 0} p_r$ is more preferable in this case. Then, at the dwell of the output unit when the system degenerates into two uncoupled vibratory circuits, the frequency p_r is equal to the partial frequency with the same index.

According to Eqs. (14.16), (14.17), the slowly varying "intrinsic" frequencies for the analyzed model are the roots of the following biquadratic equation:

$$p^4 - \{k_1^2 + k_2^2[1 + \mu^2\Pi'^2(\varphi_0)]\}p^2 + k_1^2k_2^2 = 0 \qquad (14.18)$$

The free term in Eq. (14.18) does not depend on time and, thus, based on one of the Vieta theorems $p_1(t)p_2(t) = \text{const}$. This means that the minimum of one frequency corresponds to the maximum of the other.

Figure 14.4 shows typical plots of frequency variations $p_1(\varphi_0)$, $p_2(\varphi_0)$ for two combinations of partial frequencies: $k_1 = 20$, $k_2 = 30$ (mode 1, Fig. 14.4a), and $k_1 = k_2 = 30$ (mode 2, Fig. 14.4b). Here and hereinafter, dimensionless (normalized) frequencies that correspond to the dimensionless time $\varphi_0 = \omega_0 t$ are used. In this case, a three-period structure of the motion law was used in the forward trace (speeding up, the section of uniform motion, running out) when accelerations varied following the "modified trapezium" law; with the return trace the harmonic law of acceleration variations was used [2].

Note that at similar partial frequencies there appear regions with an increased density of the frequency spectrum where the intensity of the energy exchange between vibration forms drastically increases. For the fixed value $f = \mu\Pi' < 1$, the frequencies are the closest at $\kappa = k_2/k_1 = \sqrt{1-f^2}/(1+f^2)$. At this value of f it follows that $p_2/p_1 = \sqrt{(1-f)/(1+f)}$.

Suppose when $\varphi_0 = 0$ the subsystem of the output unit receives the pulsed perturbation $(\dot{q}_2(0) \neq 0)$. Figure 14.5 ((a) is mode 1, (b) is mode 2) shows the dependencies for $q_1(t)$ (solid line) and $q_2(t)$ (dashed line) plotted on the basis of the computer simulation of the set of equations (14.2). For clarity, the plots of the dissipative terms in set (14.2) are omitted. At $\Pi'(0) = 0$, the subsystems are not

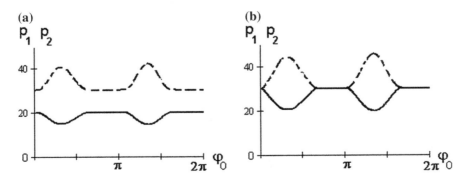

Fig. 14.4 Typical plots of frequency variations

(a)   (b)

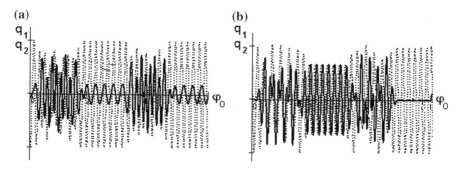

Fig. 14.5 Energy exchange between subsystems

interrelated and, thus, the initial energy store completely "belongs" to subsystem 2 (the q_2 coordinate). At $\Pi'(\varphi) \neq 0$, the energy is redistributed between both subsystems. When the partial frequencies are equal $(k_1 = k_2)$ the initial level of amplitudes is restored almost completely at the end of the kinematic cycle.

Figure 14.6a plots the variations of the normalized total energy E in both subsystems (curve 1) and the initial energy E_0 (line 2) for mode 2. On the whole, the operation of gyroscopic forces arising with variable parameters is almost zero when the vibrations are close to harmonic, and a nearly complete exchange of the vibratory energy between the subsystems is observed. As distinct from the case considered in [9] (Fig. 14.1), the cyclic system plays the function of a "trap" only partially, namely within the kinematic cycle. This is observed most clearly at the sections of uniform motion and partially at intermediate dwell.

The plots of variations in the total energy in Fig. 14.6b, c correspond to the harmonic law of motion that is widely used in engineering in which $\Pi'(\varphi_0) = r_0 \sin \varphi_0$. At similar partial frequencies (Fig. 14.6b, mode 3: $k_1 = 23$; $k_2 = 27$; $\mu^2 = J_2/J_1 = 1$) the initial energy level is not restored within one kinematic cycle. In this case, the energy accumulates over a relatively long time and yields a substantial increase in the vibrational activity of the entire system. The analysis showed that this is related to formation of the beat mode when the work of gyroscopic forces vanishes only after the beat period (which often exceeds the period of the kinematic cycle $\tau = 2\pi/\omega_0$ by one order of magnitude) is over.

Figure 14.6c plots the $E' = dE/d\varphi_0$ function which characterizes the intensity of energy variations for mode 4 ($k_1 = 10$, $k_2 = 30$, $\mu^2 = J_2/J_1 = 0.2$). This mode is interesting since, in fact, the situation considered above for the model with one degree of freedom (see Fig. 14.3) which corresponds to the limiting case when $k_2^2/k_1^2 \gg 1$ is repeated. The comparison of the E' plot with that shown in Fig. 14.3 shows that the plot of $\delta_*(\varphi_0, \mu)$ obtained at the limiting transition serves as an analogue for the envelope for high frequency vibrations that were identified when the elasticity of the output unit was taken into account. This study shows that the energy accumulated in the vibratory system in cyclic machines is mainly determined by the work performed by the external source upon implementation of nonstationary

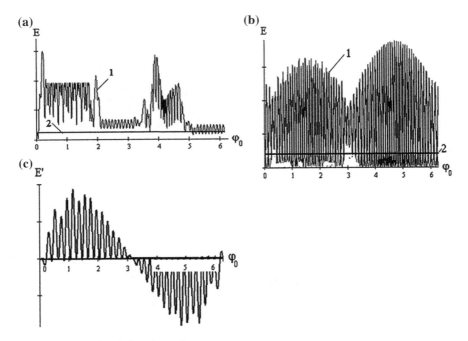

Fig. 14.6 Plots of variations in total energy

links that manifests itself most clearly in local violations of stability conditions at certain sections of the kinematic cycle. According to Eq. (14.14), these conditions can be restored by increasing the dissipative forces. The analysis showed that when dissipation is taken into account one of the subsystems (at relatively high scattering coefficients) can serve as an effective means for decreasing the level of vibrations in the second subsystem. The distribution of the energy that was transferred between the subsystems mainly depends on the ratio of partial frequencies.

In cyclic systems with a lattice structure, the role of the distribution of the energy between subsystems is even greater, since the energy factors substantially affect the transformation of non-stationary forms and spatial localization of vibrations. Besides, the variability of the parameter substantially appears not only on the kinetic energy, but also on the potential energy. This problem is partly investigated in [2, 13].

References

1. Vulfson II (1990) Kolebaniya mashin s mekhanizmami tsiklovogo deistviya (Oscillations for machines with cyclic mechanisms). Leningrad, Mashinostroenie (in Russian)
2. Vulfson I (2015) Dynamics of cyclic machines. Springer, Berlin (Translation from Russian, Politechnika, 2013)
3. Vul'fson II (2011) Phase synchronism and space localization of vibrations of cyclic machine tips with symmetrical dynamic structure. J Mach Manuf Reliab 40(1):12–18

4. Vul'fson II, Preobrazhenskaya MV (2008) Investigation of vibration modes excited upon reversal in gaps of cyclic mechanisms connected with a common actuator. J Mach Manuf Reliab 37(1):33–38
5. Vulfson I (1989) Vibroactivity of branched and ring structured mechanical drives. Hemisphere Publishing Corporation, New York (Translation from Russian, Mashinostroenie, Leningrad, 1986)
6. Indeitsev DA, Kuznetsov NG, Motygin OV et al (2007) Lokalizatsiya lineinykh voln (Linear waves localization), St. Izd. St. Petersburg University, Petersburg (in Russian)
7. Manevich LI, Mikhlin DV, Pilipchuk VN (1989) Metod normal'nykh kolebanii dlya sushchestvenno nelineinykh sistem (Normal oscillations method for strongly nonlinear systems). Nauka, Moscow (in Russian)
8. Brillouin L, Parodi M (1953) Wave Propagation in periodic structures. Dover Publications, New York
9. Kovaleva A, Manevitch L, Kosevich Y (2011) Fresnel integrals and irreversible energy transfer in an oscillatory system with time-dependent parameters. Phys Rev E 83:026602-1–026602-12
10. Kosevich YuA, Manevich LI, Manevich EL (2010) Oscillation analog of non_adiabatic Landau—Zener tunneling and possibility for creating the new type of power traps. Usp Fiz Nauk 180(12):1331–1334
11. Babakov IM (1965) Teoriya kolebanii (Oscillation theory). Nauka, Moscow (in Russian)
12. Mitropol'skii YuA (1964) Problemy asimptoticheskoi teorii nestatsionarnykh kolebanii (Problems in asymptotic theory of nonstationary jscillations). Nauka, Moscow (in Russian)
13. Vulfson II (2014) Dynamic analog to multiple-unit drives of functional members of cyclic machines that forms lattice oscillatory loops. J Mach Manuf Reliab 43(6):3–9

Chapter 15
Dynamics, Critical Speeds and Balancing of Thermoelastic Rotors

Vladimir V. Yeliseyev

Abstract Linear dynamics of a thermoelastic rotor in a rotating reference system is considered. Inertia forces change rigidity in essence. Critical regimes emerge and can be recovered by balancing adjustment. Continual and discrete mathematical models are presented.

Keywords Linear thermoelasticity · Rotating rotor · Inertia forces · Critical speeds · Balancing conditions

Introduction

A lot of works are devoted to dynamics of rotors; publications [1–4] are just a small part of works in this field. However, it is quite typical for the most of them to describe the rotor simplistically as an elastic construction, while means of theory of elasticity are used insufficiently.

Not only mechanical but also thermal loads are causing rotor deformations and its unbalancing with intensive vibration on critical speeds as a consequence.

Linear Thermoelasticity in a Rotating Reference System

We consider the following classical problem statement as a basis of modelling of rotor dynamics with thermomechanical loads [5, 6]:

V.V. Yeliseyev (✉)
Saint Petersburg State Polytechnic University, Saint Petersburg, Russia
e-mail: yeliseyev@inbox.ru

© Springer International Publishing Switzerland 2015
A. Evgrafov (ed.), *Advances in Mechanical Engineering*, Lecture Notes in Mechanical Engineering, DOI 10.1007/978-3-319-15684-2_15

Fig. 15.1 Rotor and the axes

$$\nabla \cdot \tau + \mathbf{f} = 0, \quad \tau = \frac{\partial \Pi}{\partial \varepsilon} = {}^{4}\mathbf{C} \cdot (\varepsilon - \alpha T),$$

$$\varepsilon = \nabla \mathbf{u}^{S}; \quad \mathbf{u}|_{O_1} = \mathbf{u}_0, \quad \mathbf{n} \cdot \tau|_{O_2} = \mathbf{p}, \tag{15.1}$$

where τ, ε are the deformation and strain tensors, $\mathbf{u}$ is the displacement vector, ∇ is Hamilton operator, $\mathbf{f}$ is the body force vector (with the inertial forces), Π is free energy per volume, ${}^{4}\mathbf{C}$ is the stiffness tensor of the material (4th rank), α is the given displacement on the O_1 part of the boundary, $\mathbf{p}$ is the given surface forces on the rest of the O_2 boundary.

Linear formulation (15.1) is permissible only for small displacements, strains and rotations. For applying it to the rotating rotor, it is necessary to introduce a moving reference system. The Fig. 15.1 shows the rotor rotating with the given angular speed Ω around the axis z; the Cartesian axes x_1, x_2 are rotating in the same manner.

The Cartesian axes triad x_1, x_2, z is this moving reference system in which linear formulation (15.1) is considered. Unit vectors are $\mathbf{i}_\alpha$, $\mathbf{k}$.

Two inertial forces, translational $\mathbf{f}_e$ and Coriolis $\mathbf{f}_c$, should be added to the trivial physical forces in the introduced reference system. The position vector of the point is $\mathbf{R} = \mathbf{r} + \mathbf{u}$, where vector $\mathbf{r} = \mathbf{x} + z\mathbf{K}$ is the same vector but prior to the deformation ($\mathbf{x} = x_\alpha \mathbf{i}_\alpha$; we summarize by recurring indexes). Expressions of inertial forces read

$$\mathbf{f}_e = -\rho \Omega^2 \mathbf{k} \times (\mathbf{k} \times \mathbf{R}) = \rho \Omega^2 \mathbf{R}_\perp, \quad \mathbf{R}_\perp = \mathbf{R} - R_z \mathbf{k}; \quad \mathbf{f}_c = -2\rho \Omega \mathbf{k} \times \dot{\mathbf{R}}, \tag{15.2}$$

where ρ is density. Let us note that time differentiation matches the relative motion: $\dot{\mathbf{R}} = \dot{\mathbf{u}}$ is the relative speed and $\ddot{\mathbf{R}} = \ddot{\mathbf{u}}$ is the relative acceleration. In (15.1), we substitute.

$$\widehat{\mathbf{f}} = \mathbf{f} + \rho \left(\Omega^2 (\mathbf{x} + \mathbf{u}_\perp) - 2\Omega \mathbf{k} \times \dot{\mathbf{u}} - \ddot{\mathbf{u}} \right) \tag{15.3}$$

The first term in parentheses is the centrifugal force. The physical forces $\mathbf{f}$ are, in particular, presented by the gravity and the effect of the magnetic field on the coils with current.

The system of equations and boundary conditions (15.1) is equivalent to the following variational formulation:

$$\int_V (\delta\Pi - \mathbf{f} \cdot \delta\mathbf{u})dV - \int_{O_2} \mathbf{p} \cdot \delta\mathbf{u}\, dO = 0, \quad \mathbf{u}|_{O_1} = \mathbf{u}_0. \quad (15.4)$$

This expression is the Lagrange—D'Alembert principle of virtual work. To justify this, it is sufficient to have the relations $\delta\Pi = \boldsymbol{\tau} \cdot \delta\varepsilon = \nabla \cdot (\boldsymbol{\tau} \cdot \delta\mathbf{u}) - \nabla \cdot \boldsymbol{\tau} \cdot \delta\mathbf{u}$ and the divergence theorem.

In (15.1) we have the "full system of equations." The "equations in displacements" immediately follow from it:

$$\nabla \cdot \boldsymbol{\tau}^u + \mathbf{f} + \underline{\nabla \cdot \boldsymbol{\tau}^T} = 0, \quad \boldsymbol{\tau}^u \equiv {}^4\mathbf{C} \cdot \nabla\mathbf{u}, \quad \boldsymbol{\tau}^T \equiv -{}^4\mathbf{C} \cdot \boldsymbol{\alpha}T;$$
$$\mathbf{u}|_{O_1} = \mathbf{u}_0, \quad \mathbf{n} \cdot \boldsymbol{\tau}^u|_{O_2} = \mathbf{p} \underline{- \mathbf{n} \cdot \boldsymbol{\tau}^T}. \quad (15.5)$$

The equivalents of body and surface forces are underlined. Here we have a formal similarity with the isothermal statics which allows us to apply the general theorems [5, 6]. The reciprocity theorem.

$$\int_V \mathbf{f}_1 \cdot \mathbf{u}_2 dV + \int_O \mathbf{p}_1 \cdot \mathbf{u}_2 dO = \int_V \mathbf{f}_2 \cdot \mathbf{u}_1 dV + \int_O \mathbf{p}_2 \cdot \mathbf{u}_1 dO \quad (15.6)$$

is especially useful for any of two equilibrium states. The orthogonal property of natural vibration forms is established using this theorem and ordinary differential equations (ODEs) are deduced for the principal coordinates (for the expansion in modes).

We shall clarify the form of tensors ${}^4\mathbf{C}, \boldsymbol{\alpha}$. We have $\boldsymbol{\alpha} = \alpha\mathbf{E}$ (with the identity tensor) for the isotropic material and

$$\boldsymbol{\tau} = 2\mu\left[\varepsilon + \frac{1}{1-2v}(v\varepsilon - (1+v)\alpha T)\mathbf{E}\right], \quad \varepsilon \equiv tr\varepsilon;$$
$$\Pi = \mu\left[\varepsilon \cdot \varepsilon + \frac{1}{1-2v}(v\varepsilon^2 - 2(1+v)\alpha\varepsilon T)\right], \quad (15.7)$$

where μ, v are the shear modulus and the Poisson's ratio respectively (the Young's modulus $E = 2\mu(1 + v)$); the derivative $\partial\varepsilon/\partial\varepsilon = E$.

The variational formulation (15.4) allows constructing approximate solutions by the Ritz method with the approximation given:

$$\mathbf{u} = \mathbf{U}_0 + \sum_{i=1}^N q_i(t)\varphi_i(\mathbf{r}), \quad \mathbf{U}_0|_{O_1} = \mathbf{u}_0, \quad \varphi_i|_{O_1} = 0, \quad \delta\mathbf{u} = \sum \varphi_i\delta q_i. \quad (15.8)$$

An ODE system is obtained for the function $q_i(t)$. The finite-element method is a special case of this approach.

We substitute (15.3) to (15.5):

$$\nabla \cdot \tau^u + \rho\left(\Omega^2 \mathbf{u}_\perp - 2\Omega\mathbf{k} \times \dot{\mathbf{u}} - \ddot{\mathbf{u}}\right) + \underline{\mathbf{f} + \rho\Omega^2\mathbf{x} + \nabla \cdot \tau^T} = 0;$$

$$\mathbf{u}|_{O_1} = 0, \quad \mathbf{n} \cdot \tau^u|_{O_2} = \underline{\mathbf{p} - \mathbf{n} \cdot \tau^T}. \tag{15.9}$$

This is a nonhomogeneous dynamical problem. The sources of inhomogeneity in the body and surface are underlined. The part O_1 of the body boundary is restricted to avoid "rigid-body" additions $\mathbf{u}_* + \boldsymbol{\theta}_* \times \mathbf{r}$ to the displacements.

Discretization

From the described above continual formulation, a discrete one can be derived using (15.4) and (15.8):

$$\mathbf{u} = \sum_{i=1}^N q_i(t)\varphi_i(r), \quad \varphi_i|_{O_1} = 0, \quad \varepsilon = \sum_{i=1}^N q_i\nabla\varphi_i^S, \quad \tau = \sum_{i=1}^N q_i{}^4\mathbf{C} \cdot \nabla\varphi_i^S,$$

$$\int_V \left\{ \tau \cdot \nabla\varphi_i^S - \left[\rho\left(\Omega^2\mathbf{u}_\perp - 2\Omega\mathbf{k} \times \dot{\mathbf{u}} - \ddot{\mathbf{u}}\right) + \mathbf{f} + \rho\Omega^2\mathbf{x} + \nabla \cdot \tau^T\right] \cdot \varphi_i\right\}dV$$

$$- \int_{O_2} \left(\mathbf{p} - \mathbf{n} \cdot \tau^T\right) \cdot \varphi_i dO = 0 \Rightarrow \underline{\left(C - \Omega^2\tilde{C}\right)q + \Omega G\dot{q} + M\ddot{q} = \hat{F}}, \tag{15.10}$$

where $q = (q_i)^T$ is the column of variables; the stiffness and inertial matrices

$$C = \left(\int_V \nabla\varphi_i^S \cdot {}^4\mathbf{C} \cdot \nabla\varphi_k^S dV\right), \quad \tilde{C} = \left(\int_V \rho\varphi_{i\perp} \cdot \varphi_{k\perp} dV\right),$$

$$G = -2\left(\int_V \rho\mathbf{k} \cdot \varphi_i \times \varphi_k dV\right), \quad M = \left(\int_V \rho\varphi_i \cdot \varphi_k dV\right) \tag{15.11}$$

and the column of loads

$$\hat{F} = \left(\int_V \left(\mathbf{f} + \rho\Omega^2\mathbf{x} + \nabla \cdot \tau^T\right) \cdot \varphi_i dV + \int_{O_2} \left(\mathbf{p} - \mathbf{n} \cdot \tau^T\right) \cdot \varphi_i dO\right)^T. \tag{15.12}$$

were introduced.

The equation underlined in (15.10) can be considered as the principal one in the elastic rotor dynamics. It contains symmetric and positive-definite matrices C, $\tilde{C}$, M and also an asymmetrical matrix G. The matrix of inertial forces is similar to the inertial matrix M and it defines a certain negative stiffness. The column of load F is defined by the physical forces, mass distribution and the temperature field. Based on (15.10), any non-stationary problems of rotor dynamics under the impact of power and thermal loading can be described.

To explain the following, we shall turn to the linear statics of constructions with the stiffness matrix C, the displacement column u and the load column F: $Cu = F$. The constructive solvability condition of linear algebraic system is: the right-hand side shall be orthogonal to all linearly independent solutions of the conjugate homogenous system:

$$\underline{\Psi_k^T F = 0}; \quad \Psi_k : \quad C^T \Psi = 0 \tag{15.13}$$

If $\det C \neq 0$, then $\Psi = 0$, and the solvability condition is effected for any load F. The case of $\det C = 0$ is more interesting; it is related with the dynamics of rotors. Singularity of the matrix means that a homogenous problem has a nontrivial solution.

Note that the stiffness matrix of an elastic system is symmetric: $C = C^T$. This symmetry is expressed in the reciprocity theorem. This theorem will be regularly used further on.

Conjugate Homogenous Problem and Effective Stiffness

Let us refer to (15.9) and in accordance with (15.13) consider the homogenous problem (of statics):

$$\nabla \cdot \tau^u + \rho \Omega^2 \mathbf{u}_\perp = 0; \quad \mathbf{u}|_{O_1} = 0,$$
$$\mathbf{n} \cdot \tau^u|_{O_2} = 0 \Rightarrow \Omega = \Omega_i, \quad \mathbf{u} = \Phi_i, \quad i = 1, 2, \ldots \tag{15.14}$$

We will call values Ω_i critical speeds and functions $\Phi_i(\mathbf{r})$ we will call forms. (15.14) differs from the problem of principal vibrations only by the underlined term containing $\mathbf{u}_\perp$ instead of $\mathbf{u}$.

Eigenfunctions of the problem (15.14) are similar to the modes of principal vibrations by their properties. Functions Φ_i could be considered modes in case of corresponding anisotropy of inertial properties. Orthogonality of the forms Φ_i (with normalization) is proved by means of the reciprocity theorem (15.6):

$$\int_V \rho \Phi_{i\perp} \cdot \Phi_{k\perp} dV = \delta_{ik}. \tag{15.15}$$

Applying the reciprocity Theorem (1.6) further on and formally considering (15.9) and (15.14) as two static problems, we get:

$$\int_V \left[\mathbf{f} + \rho\Omega^2\mathbf{x} + \nabla \cdot \boldsymbol{\tau}^T - \rho(2\Omega\mathbf{k} \times \dot{\mathbf{u}} + \ddot{\mathbf{u}}) \right] \cdot \boldsymbol{\Phi}_i dV$$

$$+ \int_{O_2} \left(\mathbf{p} - \mathbf{n} \cdot \boldsymbol{\tau}^T \right) \cdot \boldsymbol{\Phi}_i dO = \left(\Omega_i^2 - \Omega^2 \right) \int_V \rho\boldsymbol{\Phi}_{i\perp} \cdot \mathbf{u}_\perp dV. \tag{15.16}$$

Similar operations in the theory of vibration provide for ODEs for the principal coordinates, i.e., expansion coefficients of the vector $\mathbf{u}$ on the natural forms for calculation of arbitrary motions. Equation (15.16) is important for the following.

We have in the discrete model of a rotor

$$\left(C - \Omega_i^2\tilde{C} \right)\Phi_i = 0, \quad \left| C - \Omega_i^2\tilde{C} \right| = 0, \quad \Phi_i^T\tilde{C}\Phi_k = \delta_{ik}$$

$$\left(\hat{F} - G\dot{q} - M\ddot{q} \right)^T \Phi_i = \left(\Omega_i^2 - \Omega^2 \right)q^T\tilde{C}\Phi_i. \tag{15.17}$$

However, adequacy of this model depends on the chosen set of coordinate functions φ_i; the system of these functions should be adequately representative.

Effective rotor stiffness $\left(C - \Omega^2\tilde{C} \right)$ vanishes at the speeds Ω_i being natural values of the homogenous problems (15.14) or (15.17). This interpretation to some extent differs from the common ones. The rotor can "maintain" the load at critical speeds only dynamically, through the inertial $\left(M\ddot{q} \right)$ and gyroscopic $(G\dot{q})$ reactions.

Problem (15.14) differs from the problem of principal vibrations by the non-existent inertia of the axial motion. In this case, using Rayleigh's theorem

$$\Omega_i \geq \omega_i, \tag{15.18}$$

speeds Ω_i are no less than eigen-frequencies ω_i.

Stationary Problem and Balancing

Let's assume that forces $\mathbf{f}$, $\mathbf{p}$ and the temperature field T do not depend on time (in the rotating reference system!). At equilibrium $\Omega = \Omega_i$, we obtain from (15.16)

$$\int_V \left(\mathbf{f} + \rho\Omega^2\mathbf{x} + \nabla \cdot \boldsymbol{\tau}^T \right) \cdot \boldsymbol{\Phi}_i dV + \int_{O_2} \left(\mathbf{p} - \mathbf{n} \cdot \boldsymbol{\tau}^T \right) \cdot \boldsymbol{\Phi}_i dO = 0 \tag{15.19}$$

This condition can be the basis of the balancing theory for elastic rotors. Only with fulfillment of (15.19), the rotor will withstand the static load without transition into dynamics. If (15.19) is not fulfilled, the intensive vibrations of the rotor will take place.

It appears that practical balancing is intended for fulfillment of (15.19). Installation of a balance weight with the mass m_0 at the point positioned by the vector $\mathbf{r}_0$ means an addition of a term $m_0 \delta(\mathbf{r} - \mathbf{r}_0)$ to the function $\rho(\mathbf{r})$. Requiring fulfilment of (15.19), we will obtain

$$m_0 = \frac{\int\limits_V \left(\mathbf{f} + \rho\Omega^2\mathbf{x} + \nabla \cdot \tau^T\right) \cdot \mathbf{\Phi}_i dV + \int\limits_{O_2} \left(\mathbf{p} - \mathbf{n} \cdot \tau^T\right) \cdot \mathbf{\Phi}_i dO}{-\Omega^2(\mathbf{x} \cdot \mathbf{\Phi}_i)|_{\mathbf{r}=\mathbf{r}_o}} \tag{15.20}$$

This refers to a particular temperature field (as well as mass distribution and "static" loads $\mathbf{f}$, $\mathbf{p}$). Variation of the temperature field leads to disbalance emergence. Expression (15.20) gives exact value of the mass change caused by field variation:

$$\tilde{m}_0 = \frac{\int\limits_V \nabla \cdot \tilde{\tau}^T \cdot \mathbf{\Phi}_i dV - \int\limits_{O_2} \mathbf{n} \cdot \tilde{\tau}^T \cdot \mathbf{\Phi}_i dO}{-\Omega^2(\mathbf{x} \cdot \mathbf{\Phi}_i)|_{\mathbf{r}=\mathbf{r}_o}} = \frac{\int\limits_V \tilde{\tau}^T \cdot \nabla\mathbf{\Phi}_i^S dV}{\Omega^2(\mathbf{x} \cdot \mathbf{\Phi}_i)|_{\mathbf{r}=\mathbf{r}_o}} \quad \left(\tilde{\tau}^T = -{}^4\mathbf{C} \cdot \boldsymbol{\alpha}\tilde{T}\right) \tag{15.21}$$

Note that the denominator in (15.20) and (15.21) can turn to zero. This means that balancing is not possible and weight should be mounted at a different point.

Conclusion

The theory presented was used as a basis for a calculating method for new rotors produced at Saint Petersburg enterprises. The small relative thickness of rotors allowed to stick to the rod models with the specific for them ODEs (with respect to z) [5]. The latter were solved by means of computer math (Mathcad 14) even in case of complex irregular dependence of the section diameter upon the axial coordinate.

References

1. Goldin AC (2000) Vibration of rotor machines. Mashinostroyenie, Moscow, 344 p (in Russian)
2. Gusarov AA (1990) Dynamics and balancing of flexible rotors. Nauka, Moscow, 152 p (in Russian)
3. Kelzon AC, Cimansky YP, Yakovlev VI (1982) Dynamics of rotors in elastic bearings. Nauka, Moscow, 280 p (in Russian)

4. Poznyak EL (1980) Oscillations of rotors vibrations in technics, vol. 3. Mashinostroyenie, Moscow, pp 130–189 (in Russian)
5. Yeliseyev VV (2006) Mechanics of deformable solid bodies. Izd-vo Politechn. un-ta, SPb, 231 p (in Russian)
6. Rabotnov YN (1988) Mechanics of deformable solid bodies. Nauka, Moscow, 711 p (in Russian)

Chapter 16
Influence of Plastic Deformation
on Fatigue of Titanium Alloys

Vladimir A. Zhukov

Abstract In this work the questions are connected with research into the influence of small plastic deformations on fatigue of titanium alloys. We determined that there is a reduction of fatigue of titanium alloys after plastic deformation from 0.1 to 1 % depending on material structure and the strain resistance.

Keywords Titanium alloy · Fatigue of metals · Plastic deformation

Introduction

The sphere of titanium alloys application is extended to various branches of industry owing to high corrosion resistance under influence of corrosive environments including oxygen and the specific gravity for 1.7 times less than the constructive steels. Originally the astronautics, aviation and motor industries were the primary users of titanium alloys [1], but at present the active investigations and the practical approbations are being carried out in order to apply more prevalent application of titanium alloys for the power industry and sea oil production, as well as for hydro-construction.

It is necessary to take into account some peculiarities of the strength of titanium alloy structures together with the technological difficulties in their applications as construction material. There is a creep of titanium alloys in the climate temperature and on condition that the rated strain equals $(0.6 \ldots 0.7)\sigma_y$. Corrosion cracking resistance of middle or high durable titanium alloys was reduced substantially under influence of a water chlorine solution in the presence of a strong scratch or a crack [2]. The fatigue limit of machine parts manufactured from titanium alloys was lowered by 1.5 times because of grinding [3]. Substantial lowering fatigue resistance was founded after the polished specimens were subjected to active extension

V.A. Zhukov (✉)
Saint Petersburg State Polytechnic University, Saint Petersburg, Russia
e-mail: v.zhukoff2011@yandex.ru

© Springer International Publishing Switzerland 2015
A. Evgrafov (ed.), *Advances in Mechanical Engineering*, Lecture Notes
in Mechanical Engineering, DOI 10.1007/978-3-319-15684-2_16

or creep. Moreover, such lowering of the fatigue limit may be equal to 40 % after the creep time reaches 0.01 % of the rupture time by action of the constant stress [4]. Hence this fact indicates that fatigue fracture probability of machine parts will be increased by the action of creep stress and alternating loads simultaneously.

This work presents the results of the investigation of the small plastic deformations influence on the fatigue limit of the titanium alloys specimens.

Experimental Results

The tests were made at the symmetrical bend stress cycle. The specimens having diameter 8 or 25 mm were manufactured from plates which were subjected to bending at 20 °C. Moreover, there were tests made of the specimens subjected to creep or plastic deformations from 0.1 to 5.0 %.

Figure 16.1 illustrates the results of the testing of the specimens manufactured from material of the plates from titanium alloy BT8 after heat forging, after bending by plastic deformation of surface layer 2.5 %, also after straightening of the plates subjected to bending.

As Fig. 16.1 shows, because of bending the fatigue limit was reduced by 30 % in comparison with its initial state. Exactly, this reduction occurs from 430 to 320 MPa for 25 mm plate and from 300 to 210 MPa for 46 mm plate. In this work it was not revealed that there is a significant difference between the fatigue durability of the specimens cut of the stretched part of plates and the fatigue durability of the specimens cut of the compressed part of plates. Also, the opposite plastic straightening deformation of the plates did not assist in restoring the fatigue durability.

The size effect was investigated on the specimens with diameter of 8 and 25 mm which were cut on the plate with thickness of 100 mm from titanium alloy ПТ3В.

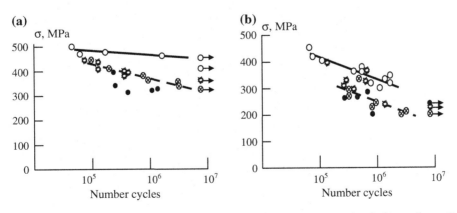

Fig. 16.1 The fatigue durability of BT8. The specimens were cut from the *thickness* plates of 25 mm (**a**) or 46 mm (**b**): ○—initial state; ✿—after bending by the plastic deformation 2.5 % (the stretched part); ⊗—after bending (the compressed part); ●—after straightening of the bended plates

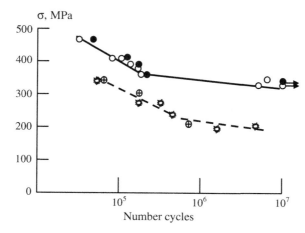

Fig. 16.2 The fatigue durability of the tensile deformation specimens from BT8: ○—initial state; ⊕—the plastic deformation 0.5 %; ✿—the plastic deformation 0.5 % and vacuum annealing; ●—the plastic deformation 0.5 % and the turning

On foundation the testing results may come to conclude that the influence of the size effect has not been discovered. This conclusion corrects for the plates subjected to bending or with initial state. Fatigue limits equal 250 MPa of initial state and 140 MPa of bending state by 2 % plastic deformation. Annealing of 870 °C in 2 h assists the fatigue durability restoring, but partially; in this case fatigue limit increases under 190 MPa.

Figure 16.2 illustrates the results of the testing of the specimens of 8 mm manufactured from material of the plates from titanium alloy BT8 with thickness 60 mm. The fatigue limit was decreased from 360 MPa of initial state to 200 MPa by the plastic tensile deformation of 0.5 %. Vacuum annealing of tensile specimens does not assist the fatigue durability restoring, even partially. However, the fatigue durability of tensile specimens was restored completely by the turning of the 0.5 mm allowance.

The plastic tensile deformation level that is the cause of the fatigue limit reduction of specimens is substantively dependent on component size, as is shown on Fig. 16.3 for plates from alloy BT8. After the plastic tensile deformation of 0.5 % the fatigue limit for 14 mm plate remains equal to the fatigue limit of the initial state. On the contrary, the fatigue limit reduction for 60 mm plate is 40 % in comparison with an initial state of equal deformation. It is possible that this effect leads to the crystal dimensions. So, there is the essential fatigue limit reduction owing to the plastic tensile deformation of only 0.1 % after high heat rolling into β-state of this alloy as shown on Fig. 16.3. It is indubitable that the plastic tensile deformation effect is dependent on resistance of plastic deformation metal. So, titanium alloy ПТ7М is not more sensitive than titanium alloy BT8, then in this case the tension yield stress of alloy BT8 is 2 times larger.

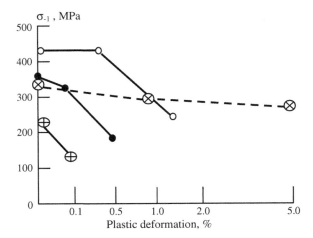

Fig. 16.3 The fatigue limit depending on the plastic deformation of specimens: ○—alloy BT8, plate 14 mm, crystal dimensions equals 0.05 mm; ●—alloy BT8, plate 60 mm, crystal dimensions equals 0.25 mm; ⊕—alloy BT8 rolling into β-state, crystal dimensions equals 1.5 mm; ⊗—alloy ПТ7М, bar ∅ 16 mm

Crystal dimension's influence was revealed in the testing of specimens subjected to turning after the plastic tensile deformation. Such as, the specimens from plate 14 mm of alloy BT8 with the crystal dimensions about 0.05 mm were deformed at 1 % and polished at the 0.05 mm allowance. The fatigue limit of those specimens increases from 240 to 450 MPa in comparison with 420 MPa for initial state. The fatigue limit of the specimens from plate 60 mm of alloy BT8 with the crystal dimensions about 0.25 mm was restored to 85 % by removing 0.5 mm allowance. However, removing allowance from 3 to 5 mm did not assist the fatigue durability restoring of material of the plates rolling into β-state of this alloy.

Negative influence of structural injury which appears by plastic deformation to be about 1 % of titanium alloys may be not removal or removal but partially by the subsequent annealing. This shows that these injuries are not removal and may be supposed as microscopic cracks.

Discussion

It is considered that dislocation sliding is the basic principle of a plastic deformation of titanium alloys by an active deformation at temperatures below the crystallization temperature. Assume that the stress estimated is the value τn in the vanguard of the group of n dislocation stopped by a crystal boundary or other structural defects, where τ is the slipping stress by the external load. In this case the microscopic crack generating criterion may be written as $\sigma_f = \tau n_f$, where σ_f is the theoretical durability

Fig. 16.4 The surface specimen injuries after tension at 1.5 % deformation; ×500

at tensile stress, n_f is the number stopped dislocation. It is considered that σ_f is equal about $0.1E$, where E is the Young's modulus. Thus the value of σ_f exceeds the value of $\sigma = \tau$, which is revealed by the external tensile or clench load. This is the explanation of the submicroscopic cracks appearance at both tension and compression. A dislocation flow assists an increase of the appeared cavity from a submicroscopic size to microscopic size, which may be observed at optical microscope (Fig. 16.4) and compared with other structures of an alloy in sizes.

The fatigue durability reduction at the small plastic deformation of other alloys specimens has been discovered, for example, construction steels, copper and copper alloys. It is possible that significant reduction of the fatigue limit of improved durability titanium alloys may be caused by alloyed components, which form distortion of the crystal lattice of titanium. In result, the dislocation sliding is concentrated in the basic plane of the hexagonal crystal lattice of titanium. It is possible that the high localization of the dislocation sliding is cause of the accelerated evolution of the structural injury from submicroscopic size to microscopic size by a small plastic deformation. According to this model the fatigue durability reduction is displayed by fatigue testing of the high durability titanium alloys with high crystal dimensions.

References

1. Filippov GA, Licyansky AC, Nazarov OI, Tomkov GP (2008) The tendency of perfection of high-speed stream engines for nuclear power stations. Energ Mach Equip, Powermachinebuilding, Saint Petersburg (3):3–12
2. Brown BF (1966) Material research and standards 6:3–129
3. Cuznetchov ND (1988) Guarantee of reliability of the contemporary aviation motors. The problems of reliability and resource into mechanical engineering. Science, Moscow, pp 51–86
4. Zhukov VA, Marinetch TC (1967) The definition of the plastic deformation influence on the material injury by the alteration of the fatigue durability The durability of metals at alternating loads. Editor-in-chief B.S. Ivanova, Moscow, pp 77–82